The Systems Psychodynamics of Organizations

THE SYSTEMS PSYCHODYNAMICS OF ORGANIZATIONS

INTEGRATING THE GROUP RELATIONS APPROACH, PSYCHOANALYTIC, AND OPEN SYSTEMS PERSPECTIVES

EDITED BY LAURENCE GOULD, LIONEL F. STAPLEY, & MARK STEIN

Routledge
Taylor & Francis Group

LONDON AND NEW YORK

First published in 2001

Published 2006 by Karnac Books Ltd.

Published 2018 by Routledge
2 Park Square, Milton Park, Abingdon, Oxon OX14 4RN
52 Vanderbilt Avenue, New York, NY 10017

Routledge is an imprint of the Taylor & Francis Group, an informa business

British Library Cataloguing in Publication Data
A C.I.P. for this book is available from the British Library

The systems psychodynamics of organizations: integrating the group relations approach, psychoanalytic, and open systems perspectives: contributions in honor of Eric J. Miller/edited by Laurence J. Gould, Lionel F. Stapley, Mark Stein.
    p. cm.
    Includes bibliographical references and index.
    ISBN: 185575441X
1. Organizational change. 2. Decision-making, Group. 3. Organization. I. Gould, Laurence J. II. Stapley, Lionel. III. Stein, Mark.
HD58.8. S948 2001
158.7–dc21
                                  2001035434

ISBN 13: 978-1-85575-441-6 (pbk)

Editors and Contibutors

Wesley Carr, M.A., Ph.D., is Dean of Westminster, England.

Rina Bar-Lev Elieli, M.A., Psy.D., is the Founder and Co-director of the Program in Organizational Consultation and Development: A Psychoanalytic and Open Systems Perspective at the Sigmund Freud Center of the Hebrew University in Jerusalem. She is also a training analyst and faculty member at the Israel Psychoanalytic Institute and Society, a faculty member of the Tel Aviv University Program of Psychotherapy, and a member of OFEK (The Israel Association for the Study of Group and Organizational Processes). Her professional work combines consultative, clinical, and teaching activities.

H. Shmuel Erlich, Ph.D., is Sigmund Freud Professor of Psychoanalysis and Director of the Sigmund Freud Center for Study and Research in Psychoanalysis, the Hebrew University of Jerusalem. He is also a training analyst and faculty member at the Israel Psychoanalytic Institute and Society, a faculty member of the Program in Organizational Consultation and Development: A Psychoanalytic and Open Systems Perspective at the Sigmund Freud Center, and a founder and board member of OFEK (The Israel Association for the Study of Group and Organizational Processes).

Laurence J. Gould, Ph.D., is a professor of Psychology in the Clinical Psychology Doctoral Program at the City University of New York, the Founding Director of the Socio-Psychoanalytic Training Program in Organizational Development and Consultation at the Institute for Psychoanalytic Training and Research (IPTAR), the Co-director of the Program in Organizational Development and Consultation: A Psychoanalytic and Open Systems Perspective at the Sigmund Freud Center of the Hebrew University in Jerusalem, a member of OFEK (The Israel Association for the Study of Group and Organizational Processes), and Co-editor of *Organisational and Social Dynamics: An International Journal Integrating Psychoanalytic, Systemic and Group Relations Perspectives*. He works as a psychoanalyst and organizational consultant, and is a founding member of the International Society for the Psychoanalytic Study of Organizations.

James Krantz, Ph.D., is a Principal with the TRIAD Consulting Group in New York City. He is a founding member of the International Society for the Psychoanalytic Study of Organizations and a Fellow of the A.K. Rice Institute.

Anton Obholzer, BSc, MB, ChB, DPM, FRCPsych, is a psychiatrist and a Fellow of the Royal College of Psychiatry, and a psychoanalyst. He is also Chief Executive of the Tavistock & Portman NHS Trust, Chairman of Tavistock Clinic's Consulting to Institutions Workshop, a Senior Consultant at the Tavistock Consultancy Service, and the Associate Director of the Tavistock Institute of Human Relations Group Relations Training Programme.

Edward Shapiro, M.D., is the Medical Director/CEO of the Austen Riggs Center in Stockbridge, Massachusetts, a faculty member of the Boston Psychoanalytic Institute, and an Associate Clinical Professor of Psychiatry at Harvard Medical School.

Ralph Stacy, B.Comm, MSc. (Econ.), Ph.D. (Econ.), is a Professor of Management and Director of the Complexity and Management Centre at the Business School of the University of Hertfordshire in the United Kingdom, as well as a Visiting Fellow at Roffey Park Management Institute.

Lionel F. Stapley, MSc., Ph.D., is an organisational consultant and the Director of OPUS (An Organisation for Promoting Understanding in Society), an educational charity which seeks to encourage the study of conscious and unconscious processes in society and institutions within it. He is a member of the International Society for the Psychoanalytic Study of Organisations, a Fellow of the Institute of Personnel and Development (FIPD), a Fellow of the Institute of Management (FIM), as well as a faculty member in the Socio-Psychoanalytic Training Program in Organizational Development and Consultation at the Institute for Psychoanalytic Training and Research (IPTAR).

Mark Stein, Ph.D., is PriceWaterhouseCoopers Fellow at the Centre for Analysis of Risk and Regulation (CARR), London School of Economics. Previously, he held posts at South Bank University, Brunel University, and the Tavistock Institute. He is also a visiting tutor at the Tavistock Clinic and an associate of OPUS (An Organisation for Promoting Understanding in Society).

Kathleen Pogue White, Ph.D., is a psychologist-psychoanalyst. She is a faculty member at the William Alanson White Psychoanalytic Institute's Program in Organization Development and Consultation, an adjunct Associate Professor in the Clinical Psychology Doctoral Programs at Columbia University and the City University of New York, and faculty member of the Annie E. Casey Foundation Fellowship Program. She is also a Fellow of the A.K. Rice Institute and the Australian Institute of Socio-Analysis.

For Eric, whose work and ideals have been an inspiration to us and untold others

Contents

Introduction

LAURENCE J. GOULD

After a long and highly creative period as Director of the Group Relations Training Programme at the Tavistock Institute, Dr. Eric J. Miller retired from this position in 1998. This book, and a forthcoming volume, *Experiential Learning in Organizations: Applications of the Group Relations Approach* (Gould et al., in press), seeks to pay tribute to Eric's extraordinary career. However, the reader should not expect a *Festschrift* in the traditional sense. Rather, this volume is restricted to honoring Eric through one particular aspect of his long and varied career. All of his work shares a common theme; namely, his commitment to and his development of a learning-from-experience model of education about group and organizational life. This volume, directed at the theme of "systems psychodynamics" reflects his generativity and creativity in developing the bases of this important theoretical and applied domain. It is also an opportunity for tributes to be paid by those many people who have a relationship with Eric that has allowed them to carry his work forward. Some of these are close friends and colleagues, while others for whom his work was—and is—an inspiration, are among the many who wish to say "Thank you Eric!"

At least in a formal sense, the field of systems psychodynamics had its birth with the publication of Miller and Rice's seminal volume *Systems of Organization* (1967). While many of the elements of this perspective were already present in the pioneering work of the Tavistock Institute of Human Relations (now the Tavistock Institute), it was not until this volume was published that they were put into a systematic framework that could rightly be called an interdisciplinary field which attempted to integrate the emerging insights of groups relations theory, psychoanalysis, and open systems theory. It is also fair to say that this is still an emerging field of social science, the boundaries of which are continually being refined and redefined. The purpose of this volume, then, is to present the most up-to-date exemplars that currently give systems psychodynamics its shape and meaning.

We would also like to note that there are many talented colleagues throughout the world who might have been invited to contribute to this volume. It is a testimony to Eric's influence—directly and indirectly—that this number is quite large. Sadly therefore, while all could not be accommodated between the covers of this volume, we are certain that they share our thanks and gratitude for Eric's contributions.

While Eric's work in the field of group and organizational studies is wide-ranging, we felt that his work in this area represented one of the core contributions of his career. As such, the frame for this book is to show the various ways in which the boundaries of the systems psychodynamic approach[1] have been pushed out and extended to provide creative, conceptual, and applied links, with a view to better understanding organizations, including the challenges of management and leadership, and a praxis for work group and organizational education, training, and consultation.

WHAT IS SYSTEMS PSYCHODYNAMICS?

The central tenet of the systems psychodynamic perspective is contained in the conjunction of these two terms. The "systems" designation refers to the *open systems* concepts that provide the dominant framing perspective for understanding the structural aspects of an organizational system. These include its design, division of labor, levels of authority, and reporting relationships; the nature of work tasks, processes, and activities; its mission and *primary task*; and, in particular, the nature and patterning of the organization's task and *sentient* boundaries and the transactions across them. Fenichel (1946) noted, for example, that human beings create social institutions to satisfy their needs as well as to accomplish required tasks, but that institutions then become external realities, comparatively independent

---

1. The term itself has only come into common usage during the past five or six years. In earlier times it was often simply referred to as the "Tavi approach" and now also sometimes referred to as "socio-analysis."

of individuals. These affect individuals in significant emotional and psychological ways, and consequently learning about their impact can be of enormous value in shedding light on the dilemmas that members of organizations may be facing. It is perhaps not surprising that learning from experience should be of fundamental concern for those working in the systems psychodynamic tradition, with its focus on development, insight, understanding, and "deep" change. Despite this, except in the therapeutic realm itself, such learning has neither been extensively nor systematically explored in the literature.

The "psychodynamic" designation refers to *psychoanalytic perspectives* on individual experiences and mental processes (e.g., transfererence, resistances, object relations, fantasy, etc.), as well as on the experiences of unconscious group and social processes, which are simultaneously both a source and a consequence of unresolved or unrecognized organizational difficulties.

Further, the systems psychodynamic framework is specifically intended to convey the notion that the observable and structural features of an organization—even quite rational and functional ones—continually interact with its members at all levels in a manner that stimulates particular patterns of individual and group dynamic processes. In turn, such processes may determine how particular features of the organization come to be created, such as its distinctive culture; how work is conceived, organized, and managed; and how it is structured. Although, as noted, the term was not yet used at the time that *Systems of Organization* was written, this volume provides what is still probably the most recent theoretical statement regarding the integration of systemic and psychodynamic conceptions of work groups and organizations. Those who find the present volume of interest would be well repaid to return to the source.

On a psychodynamic level, a central feature of the systems psychodynamic view posits the existence of primitive anxieties (of a persecutory and depressive nature), and the mobilization of *social defense systems* against them (e.g., Jaques 1955, Menzies 1960). It is useful to emphasize in this context that the idea of *social defenses* was one of the earliest major organization constructs defining the Tavistock's unique approach to organizational life. It provides an obvious analog to the conception of individual defenses which are central to psychoanalytic theory and practice. The operation of such defenses—like their analogs in an individual's psychological functioning—are conceptualized as either impeding or facilitating task performance and responses to, and readiness for change and new learning. Interventions based on this perspective typically involve understanding, interpreting, and *working through* such collective defenses, which hopefully result in enlarging the organization's capacity to develop task-appropriate adaptations that include a more rational distribution of authority, and clearer role and boundary definitions, together with their management and regulation.

Having outlined some major strands of the systems psychodynamic perspective, it may be useful to sum up by reviewing its essential features. As Armstrong notes (1995), one perspective regarding the conjunction of *systems* and *psychodynamics* may be to liken it to Bion's "reversible perspective" in which a phenomena can be viewed in a manner similar to a figure/ground illusion. Or again, he continues, ". . . is it a . . . provisional way of pointing to or naming something new, neither 'psychoanalytic' nor 'systemic', but psychoanalytic and systemic'; and an emergent but not yet fully disclosed third?" (p. 1). In light of the material provided above, both views would seem to have a place, and are by no means mutually exclusive. In the "reversible perspective" sense, consultations often do shift their focus from level to level, and from one type of analysis to another, depending on both the state of the diagnosis or consultation at any given time, as well as on the nature of the presenting issues. With regard to what is "new," as Armstrong suggests it is precisely the conjunction itself that creates the emergent, but not yet fully articulated field of systems psychodynamics. In this vein it may be argued that a systems psychodynamic perspective implies working simultaneously from "the inside out" and "the outside in," with neither perspective being privileged.

THE CONCEPTUAL ORIGINS OF THE SYSTEMS PSYCHODYNAMIC PERSPECTIVE

Psychoanalysis

From its earliest days, psychoanalysis has been interested in the nature of group and organizational processes. For example, in *Group Psychology and the Analysis of the Ego* (1921), Freud linked certain dynamic aspects of the church and the army to his earlier hypotheses regarding the origins of social process and social structure—namely, in his analysis of the primal horde (1913). Indeed, in his very first sentence he says: "The contrast between individual and social or group psychology, which at first glance may seem to be full of significance, loses a great deal of its sharpness when it is examined more closely." However, despite this early interest in group psychology, and some additions in Freud's (1927, 1930, 1939) later "sociological works," neither he nor his colleagues carried this line of theorizing much further. It should also be noted that while there have been substantial advances since that time, with current group and organizational professionals who practice in this tradition utilizing a wide variety of psychoanalytic concepts, Bion's work and the Kleinian concepts that it is based on, in part, are still the touchstone.

Group Relations Theory and Training

However, the explicit psychoanalytic origins of a systematic approach to group life was not to happen until the publication of *Experiences in Groups* (1961) in which Bion put forward a theory of group processes, based largely on developments in object relations theory pioneered by Melanie Klein (1928, 1935, 1940, 1945, 1946, 1948, 1957) and her colleagues. The essential element of Bion's theory of group life was to differentiate between behaviors and activities geared toward rational task performance (the *work* [W] *group*), and those geared to emotional needs and anxieties (the *basic assumption* [ba] groups—*fight/flight* [baF], *dependency* [baD], and *pairing* [baP]). Following Klein, he viewed these as manifestations of experiences and unconscious fantasies originating in infancy. In addition to noting the importance of Klein's general theory of development centering on the *paranoid/schizoid* and *depressive positions* (see Gould 1997), Bion also explicitly set out to articulate the relevance of other central Kleinian concepts for an understanding of the human group, including: *projective identification, splitting, psychotic anxiety, symbol formation, schizoid mechanisms, and part-objects.* It was this extraordinarily influential work that provided the major psychoanalytic underpinnings of what came to be known as the Tavistock approach to the theory and practice of both group relations training and organizational theory and consultation.

The essentials of the group relations conference (e.g., Miller 1989, 1990a, 1990b, Rice 1965, Gould, in press) are as follows: It is a conference designed as a temporary educational institution, comprised of both members and staff, which can be studied experientially as it forms, evolves, and comes to an end. As described below, the conference provides various settings for the here and now study of the relatedness of individual, group, and organization. The primary task of the conference focuses on the theme of "Authority, Leadership, and Organization," which is generally defined in terms such as "to provide opportunities to study the exercise of authority in the context of inter-personal, inter-group and institutional relations within the Conference Institution."

As will be seen, the primary task is not to directly contribute to nor explicate theory. Its aim is psycho-educational in that it provides members with opportunities to learn about their own involvement in the dynamics of the conference, with a special focus on learning about the nature of authority and the problems encountered in its exercise. As Miller (1989) states, "More generally, the aim is to enable 'the individual to develop greater maturity in understanding and managing the boundary between his own inner world and the realities of his external environment' . . . in other words, to struggle to exercise one's own authority, to manage oneself in role and to become less of a captive of group and organizational processes" (p. 8).

Open Systems Theory

One of the early origins of *open systems* theory was significantly influenced by the work of Lewin (1947), who recognized the importance of studying groups as a whole. For example, Lewin (1947) observed that, "In the social as in the physical field the structural properties of a dynamic whole are different from the structural properties of sub-parts. Both sets of properties have to be investigated . . ." (p. 8).

However, the more contemporary version of open systems theory originated with Miller's (1959) important paper on boundary differentiation and Rice's early work at the Tavistock Institute, where he was a member of the team that conducted its first major project—a consultancy to the Glacier Metal Company, headed up by Elliot Jacques. Among the papers that resulted from this work, several were seminal. The first, of which Rice was the senior author with Hill and Trist, was titled *The Presentation of Labour Turnover as a Social Process* (1950). This paper described what was then a new approach to the study of labor turnover in which the organization was conceptualized as an *open system*. In this paper Rice attempted to show how the passage of members through it, from joining to leaving, could be identified as "a continuous, mathematically coherent process, analogous to organic metabolism." It also showed the importance of subsystems, each with its unique character and *primary task*, and each related to the enterprise as a whole.

In a further publication, Rice adds yet another major theme to his conceptual repertoire—the concept of *socio-technical system*. This concept had already been quite well developed in the context of the Tavistock's studies of work organization in coal mining that was led by Trist. Rice tested it experimentally in the loom-sheds of a textile company in Ahmedabad, India. The publication of this work in his book titled *Productivity and Social Organization: The Ahmedabad Experiment* (1958), brought Rice, for the first time, international recognition. In the first part of the book, which is more widely known than the second part, he introduced the concept of primary task in relation to open systems theory. This concept was to become pivotal in the development of the Tavistock Institute's theory-building enterprise. Finally, the nature and patterning of the organization's task and *sentient* boundaries and the transactions across them was further developed and more formally articulated by himself and Miller in their book *Systems of Organizations* (1967).

A further critical aspect of boundaries is that they can be viewed as the major postulate or derivative of open systems theory (i.e., von Bertalanffy 1950a,b). As Miller (1989) explains, "The notion of the open system made it possible to look simultaneously both at the relationship between social and technical and also at the relationships between the part and the whole, the whole and the environment—including, of course, the individual and the group, and the group and the organization" (p. 11). As noted, systems theory provides a key connecting concept, that of boundary. To quote from Miller again, "The existence and survival of any human system depends upon continuous interchange with its environment, whether of materials, people, information, ideas, values, or fantasies. The boundary across which these 'commodities' flow in and out both separates any given system from, and links it to, its environment. It marks a discontinuity between the task of that particular system and the tasks of the related systems with which it transacts" (p.11).

"Because these relations are never stable and static, and because the behavior and identity of the system are subject to continual renegotiation and redefinition, the system boundary is best conceived not as a line but as a region. That region is the location of those roles and activities that are concerned with mediating relations between inside and outside. In organizations and groups this is the function of leadership; in individuals it is the ego function" (p. 11). Boundaries are a critical concept in all aspects of systems psychodynamic theory and practice. These include: time boundaries; task boundaries; territorial boundaries; role boundaries between staff and superiors and subordinates, and between the different roles that might be taken by the same person at different times; between person and role; and between the inner world of the individual, or the group, and the external environment.

APPLYING THE SYSTEMS PSYCHODYNAMIC PERSPECTIVE

There is a wide spectrum of consultation practice encompassed by the systems psychodynamic perspective. Bain (1982), for example, provides a quite useful taxonomy. He notes that the organizational consultant actually engages in an analysis of the interrelationships of some or all of the following: boundaries; roles and role configurations; structure and organizational design; and work culture and group processes. Of these, the analysis of work culture and group processes is most closely related to group relations training conferences, the aim of which is to explore the psychodynamics of the various groups that comprise the institution, one's participation in them, and the dynamic aspects of the conference institution viewed structurally and as an emergent culture. Such practice may be viewed as psychoanalytic process consultation, or in Bion's (1961) terms "therapy of the group" (as distinct from "group therapy").

A consultant working on these aspects of the organization would be alert to, and selectively interpret the covert and dynamic aspects of the client organization and the work groups that comprise it, often with a focus on relatedness and how authority is psychologically distributed, exercised, and enacted, in contrast to how it is formally invested. This work would include a consideration of attitudes, beliefs, fantasies, core anxieties, social defenses, patterns of relationships and collaboration, and how these in turn may influence task performance. Further, in a manner similar to group relations training conferences, the consultation work would focus on how a variety of unwanted feelings and experiences are split off and projected onto particular individuals and groups that *carry them*—that is, their *process roles* as distinct from their formally sanctioned roles—on behalf of the organization. There are also a variety of consultations to individuals, focusing specifically on how they take up their work roles, that draw on psychoanalytic theories and methods, and an understanding of open systems. The basic approach typically takes the form of a time-limited series of meetings (Reed 1976), which generically is often referred to as "organizational role consultation." In this work, consultants may utilize a variety of assessment and intervention methods including projective or semi-projective measures (e.g., the production and analysis of "mental maps" [Gould 1987]), the solicitation and interpretation of dream material and fantasies stimulated by the social surround of the organization, and supportive/interpretive interventions. The overarching aims of such work, as articulated by Armstrong (1995), are to ". . . [disclose] meaning: to [introduce] the client . . . to the organization-in-himself and himself-in-the-organization" (p. 8). He goes on to say that it provides ". . . a repertoire of reflected experience which the client can draw on in thinking through [the] dilemmas and challenges from the perspective of his or her own role: assessing risks, foreshadowing responses, modulating actions, communicating goals, containing anxieties [and] releasing energies" (p. 8).

This form of consultation can be especially useful in situations in which a key executive or manager in the client system has either taken up a new role, needs to reassess role performance in light of changing organizational circumstances, or is experiencing chronic difficulties in functioning effectively. With regard to the latter, such consultations may also be useful in helping an individual recognize major characterological issues and tendencies which impinge on role performance (e.g., Kernberg 1979), if they are not so severe as to be uninfluenced by insight, which can be translated into more adequate self-management (see Lawrence 1979). However, it is important to emphasize that the organizational role consultation process is not a form of counseling or psychotherapy for problem managers; rather, it is a developmentally focused, psychoeducational process for key staff, at any level, whose roles are critical to the organization's functioning.

Finally, Bain reminds us that, as noted above, one of the earliest developments in the area of systems psychodynamic organizational theory was the pioneering work of Menzies on the relationship between task, process, and structure. Starting with her classic paper on the organization of a nursing service in a general hospital, Menzies (1960) argued, for example, that instrumental structural arrangements, such as a seemingly reasonable and appropriate division of labor, often contain elements of a social defense system that, in the case of the nurses she studied, functioned to reduce the stresses of sustained contact with seriously ill patients. Subsequently, many theorists and practitioners (e.g., Hirschhorn 1990, Hirschhorn and Young 1993; Diamond 1993) found abundant evidence to support the idea that unconscious anxieties are often reflected in organizational structure and design, which function to defend against them. Therefore, analyzing the social defense aspects of organizational structure and its relationship to task and process is yet another critical feature of systems psychodynamic practice.

The issues outlined above will be further referred to and expanded upon in the chapters that follow, in what is intended to be an authoritative source book on the learning and creative application of the systems psychodynamic perspective. In a quite direct sense these chapters, in large measure, define the field by presenting key concepts, models, and social methodologies that derive from it, together with their theoretical and conceptual underpinnings in psychoanalysis, group relations, and open systems theory. However, each chapter is self-contained and reflects a particular aspect of how the systems psychodynamic perspective can be applied to working with a wide spectrum of contemporary educational, community, and organizational concerns, ranging from the individual to large systems. The chapters cover the broad issues of authority, unconscious group processes, leadership, organizational culture, and social defenses, focusing on dilemmas such as inter-group conflict, organizational change and transition, and relationships with the external environment, as well as applications in particular settings.

The first chapter relates to application work with individuals in organizations. Kathleen Pogue White outlines the major dilemmas of working with an individual client where both personal and organizational issues are critical, and how to think about the boundary between psychotherapy and organizational role consultation. Through an extended case example she details her struggles to provide both effective treatment and consultation to her patient/client and what lessons can be learned from such struggles.

In Chapter 2, Wesley Carr examines the interrelated issues of dependency and dependence and their relationship to power and authority. Writing as he does from the perspective of a church leader who has first-hand experience of followers being at times overtly dependent on him and his colleagues, he is able to provide an insightful view of these issues. Through a balanced linking of theory and application, he helps us to better understand the complex nature of these phenomena. This is particularly so regarding the nature of authority as a "negotiated concept," and the dynamic context in which it is being exercised.

The third chapter, by Rina Bar-Lev Elieli, presents her consultation work to a clinic that provides therapeutic services to children who, as a consequence of immigration, having been orphaned, or subject to family breakup, do not live with their parents. Using a large group consultation to the entire staff team, she shows some of the complex processes that were being experienced within and on the boundary of the institution at a time of major flux. These include an identification with the client at a time when the "parent" funding agency of the clinic was being changed. Following a detailed description and analysis of the case, she proposes the interesting hypothesis that being able to contain the notion of a "small group" within oneself may be of considerable help to those working in a large-group setting.

As the title suggests ("Complexity at the 'Edge' of the Basic-Assumption Group"), Chapter 4 by Ralph Stacey serves as a timely reminder that we need to consider new approaches that may challenge or complement our traditional thinking. He writes from the unique position of "complexity theory" as applied to organizations; he has also been an active participant in group relations training activities. Here, Stacey seeks to view the group relations approach to consultation from a complexity perspective, with a particular emphasis on basic assumption activity.

The fifth chapter, "Enemies Within and Without: Paranoia and Regression in Groups and Organizations," by H. Shmuel Erlich, examines the issues of paranoia and enmity that crucially affect our everyday living and surviving, both as citizens and as members of institutions. From the combined vantage points of psychoanalysis and group relations theory, and using a variety of applications, he deftly provides us with an interesting and informative means of gaining a deeper understanding of these difficult but all-pervading issues. None of us can, directly or indirectly, escape from the consequences of institutional and societal dynamics that result in the creation of an "enemy" seen as some sort of crazy individual. However, as Erlich convincingly shows, those so identified are not "lonely, deranged, withdrawn, hallucinatory, clinically delusional, mentally ill persons . . . they are more likely to be well-known persons in critical leadership positions who are pushed to extreme limits by group, institutional, or social and political processes."

In Chapter 6, James Krantz discusses the efforts of organizations to bring about major changes in the way they function, and the fact that such changes, while necessary, are also profoundly disruptive both to the organizations and to the people functioning within them. In this connection he puts forward the hypothesis that major organizational change efforts pose great psychic challenges to their members and require, in response, distinctive conditions in order to adequately contain the profound anxieties evoked by such upheavals; and further, if these conditions are absent, efforts to change are likely to fail.

The remaining three chapters relate to working with a total system. In Chapter 7, Lionel F. Stapley provides the reader with an insight into one of the most frequently cited yet seldom understood phenomena of organizations: organizational culture. As he notes, "if we are to work in organizations, and particularly if we are attempting to bring about organizational change, it is important that we have a means of studying and understanding this phenomenon." This chapter examines the influence of organizational culture; identifies some of the problems that it creates for people working in organizations; and demonstrates a practical means of analyzing this elusive phenomenon. The approach he takes is to outline the theory and then apply it to a large system intervention in a health care system as a means of showing how a systems psychodynamic perspective can be helpful—even essential—in providing the desired understanding necessary to guide a consultation effort.

In Chapter 8, Edward R. Shapiro recounts the experience of taking the leadership of a small psychiatric hospital with a long, illustrious history. Using concepts that derive from the systems psychodynamic framework, he explores the challenges and dilemmas of maintaining this institution's links with its valued tradition and at the same time leading it in a changing world. In doing so, Shapiro discusses many topics, including the reassessment of the use of groups and how they function, the gradual development of a shared recognition of institutional dynamics through creating a culture of negotiated interpretation, and the process of using a consultation approach as a critical element in his role as a manager and leader.

In Chapter 9, Anton Obholzer alerts us to the core issues of institutional leadership: creating an appropriate institutional vision, managing the boundary "osmotic process" in the context of the strategic timetable and a plan for changing the organization, and creating a containing environment conducive to creativity and thoughtfulness. In this connection he emphasizes the necessity for leadership to understand and work at the task of bringing into awareness defensive, bureaucratic time-wasting activities that, taken together, form a network of resistances to change. He also provides an excellent overview of the corresponding issues of followership, and argues that for an organization to be creative it requires followership to be an active process of engagement and participation, without which realistic leadership is not possible.

This book will be of interest to all who are concerned with the theory and practice of working with groups and organizations from whatever perspective, as well as those who have a serious general interest in the area: MBA and other management students, managers, consultants, personnel and training professionals, researchers in group and organizational behavior, and academics. It is stressed, however, that it is not the intention of this book to solely "preach to the converted." Rather, it is aimed at providing an overview of the systems psychodynamic approach together with an array of detailed examples for those who have virtually no background in this field. It will appeal, therefore, to a wide variety of professionals and students who are broadly interested in group and organizational psychology, as well as those who are already conversant with the theories and practice of systems psychodynamics and its applications.

REFERENCES

Armstrong, D. (1995). *The analytic object in organizational work*. Paper presented at the annual meeting of the International Society for the Psychoanalytic Study of Organizations, London.

Bain, A. (1982). The Baric experiment. *Occasional Paper No. 4*. London: Tavistock Institute of Human Relations.

Bertalanffy, L. von (1950a). The theory of open systems in physics and biology. *Science* 3:23–29.

——— (1950b). An outline of general systems theory. *British Journal of Philosophy of Science* 1:134–165.

Bion, W. R. (1961). *Experiences in Groups*. New York: Basic Books.

Diamond, M. (1993). Bureaucracy as externalized self-system: a view from the psychological interior. In *The Psychodynamics of Organizations*, ed. L. Hirschhorn and C. K. Barnett. Philadelphia: Temple University Press.

Fenichel, O. (1946). *The Psychoanalytic Theory of Neuroses*. London: Heinemann.

Freud, S. (1913). Totem and taboo. *Standard Edition of the Complete Psychological Works of Sigmund Freud*. Vol. 13. London: The Hogarth Press, 1955. 1–161.

——— (1921). Group Psychology and the Analysis of the Ego. *Standard Edition* 18:69.

——— (1927). The future of an illusion. *Standard Edition* 21:5–56.

————— (1930). Civilization and its discontents. *Standard Edition* 21:64–145.

————— (1939). Moses and monotheism: three essays. *Standard Edition* 23:7–137.

Gould, L. J. (1987). *A methodology for assessing internal working models of the organization: applications to management and organizational development programs.* Paper presented at the annual meeting of the International Society for the Psychoanalytic Study of Organizations, New York City.

————— (1997). Correspondences between Bion's basic assumption theory and Klein's developmental positions: an outline. *Free Associations* 7:15–30.

————— (in press). Fraternal disciplines: group relations training and system psychodynamic consultation. In *Experiential Learning in Organizations: Applications of the Group Relations Approach*, ed. L. J. Gould, L. F. Stapely, and M. Stein. Madison, CT: Psychosocial Press.

Gould, L. J., Stapely, L. F., and Stein, M. (in press). *Experiential Learning in Organizations: Applications of the Group Relations Approach.* Madison, CT: Psychosocial Press.

Hirschhorn, L. (1990). *The Workplace Within: The Psychodynamics of Organizational Life.* Cambridge, MA: The MIT Press.

Hirschhorn, L., and Young, D. R. (1993). The psychodynamics of safety: a case study of an oil refinery. In *The Psychodynamics of Organizations*, ed. L. Hirschhorn and C. K. Barnett. Philadelphia: Temple University Press.

Jaques, E. (1955). Social systems as a defense against persecutory and depressive anxiety. In *New Directions in Psychoanalysis*, ed. M. Klein, P. Heimann, and R. Money-Kyrle. New York: Basic Books.

Kernberg, O. (1979). Regression in organizational leadership. *Psychiatry* 42:24–39.

Klein, M. (1928). Early stages of the Oedipus conflict. *The International Journal of Psycho-Analysis* 9:167–180.

————— (1935). A contribution to the psychogenesis of manic depressive states. *The International Journal of Psycho-Analysis* 16:145–174.

————— (1940). Mourning and its relation to manic depressive states. *The International Journal of Psycho-Analysis* 21:125–153.

————— (1945). The Oedipus complex in the light of early anxieties. *The International Journal of Psycho-Analysis* 26:11–33.

————— (1946). Notes on some schizoid mechanisms. *The International Journal of Psycho-Analysis* 27:99–110.

————— (1948). *Contributions to Psychoanalysis 1921–1945.* London: The Hogarth Press.

————— (1957). *Envy and Gratitude.* New York: Basic Books.

Lawrence, W. G. (1979). A concept for today: the management of oneself in role. In *Exploring Individual and Organizational Boundaries*, ed. W. G. Lawrence. New York: John Wiley & Sons.

Lewin, K. (1947a). Frontiers in group dynamics. *Human Relations* 1:2–38.

———— (1947b). Frontiers in group dynamics. *Human Relations* 1:143–153.

Menzies, I.E.P. (1960). A case study in the functioning of social systems as a defence against anxiety: a report on the nursing service of a general hospital. *Human Relations* 13. Also in Coleman, A. D. and Bexton, W. H., eds. (1975). *Group Relations Reader*. A. K. Rice Institute, pp. 281–312.

Miller, E. J. (1989). The Leicester model: experiential study of group and organizational processes. *Occasional Paper No. 10*. London: Tavistock Institute of Human Relations.

———— (1990a). Experiential learning in groups I: the development of the Leicester model. In *The Social Engagement of the Social Sciences: A Tavistock Anthology, Vol. 1, The Socio-Psychological Perspective*, ed. E. L. Trist and H. Murray. Philadelphia, PA: University of Pennsylvania Press/Free Association Books.

———— (1990b). Experiential learning in groups II: recent developments in dissemination and application. In *The Social Engagement of the Social Sciences: A Tavistock Anthology, Vol. 1, The Socio-Psychological Perspective*, ed. E. L. Trist and H. Murray. Philadelphia, PA: University of Pennsylvania Press/Free Association Books.

Miller, E. J., and Rice, A. K. (1967). *Systems of Organization*. London: Tavistock Publications.

Reed, B. (1976). Organizational role analysis. In *Developing Skills in Managers*, ed. C. L. Cooper. London: The Macmillan Press, pp. 89–102.

Rice, A. K. (1958). *Productivity and Social Organisation: The Ahmedabad Experiment*. London: Tavistock Publications.

———— (1965). *Learning for Leadership: Interpersonal and Intergroup Relations*. London: Tavistock Publications.

Rice, A. K., Hill, J.M.M., and Trist, E. L. (1950). The representation of labour turnover as a social process: Glacier Project–II. *Human Relations* 3:349–372.

Stapley, L. (1999). Introduction. In *Applied Experiential Learning: The Group Relations Training Approach*, ed. L. J. Gould, L. F. Stapley, and M. Stein. Madison, CT: Psychosocial Press.

Winnicott, D. W. (1965). *The Maturational Processes and the Facilitating Environment*. New York: International Universities Press.

1

Applying Learning from Experience: The Intersection of Psychoanalysis and Organizational Role Consultation

KATHLEEN POGUE WHITE

This is an experiential narrative about discovery. It is a personal narrative about the discovery of a way of learning that links both inner experience and inner models of reality to processes of evolution, development, and change. At the heart of my discovery is an understanding that when inner models of experience are made conscious, are shared, and are debated, these can become the building blocks of new learning, and that the process itself can have two significant effects: (1) the creation of changed inner models—an antecedent to personal development (Issacharoff 1979), and (2) the creation of changed shared reality—an antecedent to system development in the interpersonal or intergroup spheres of experience (De Geus 1988).

The process of becoming conscious of and sharing inner constructs of how reality works is quite an ordinary part of living and relating, akin to the processes of mutual identification from which relatedness evolves. But as a model of learning, that might be made intentional and applied in my work, it was important to my experiential discovery. It stands in contrast to what must have been my earlier picture of the learning process. On reflection, I must have thought of learning as more simply a throughput of information; either as the acquisition of information to reach a desired outcome, for instance, to obtain a professional degree or to better manage a personal project; or as receipt or delivery of information through a teaching process, again to reach a desired goal. While these parts of a learning model contain some dynamic processes—certainly unconscious identification with authors and teachers is among them—the dynamic processes themselves are not intentionally, nor explicitly used in reaching the desired outcome.

The experiential discovery of a way of learning that explicates inner experience in order to develop shared models of reality first came through a membership experience in a working conference in the Tavistock tradition. While my learning has been amplified over time through experience in the roles of psychoanalyst and organizational role consultant, the discovery of a way to learn and to apply learning began there, and has had a lasting consequence in my personal and professional development: I became aware of the links between self and system, that is, the connection between thought, behaviors, feelings, and the dilemmas encountered in organizational role—my own and others. I have since become aware of the intersection of traditions about self and system—psychoanalysis and systems theory. More specifically, I have become aware of the reciprocally informing nature of these traditions of thought in application to the consultant role that contains the two practice areas of psychoanalysis and organizational role consultation.[1]

The purpose of this paper is to trace the threads of my experiential discoveries through to some reflection about the application of the learning transfer between the two interconnected practices, and to describe the cycle of my learning that is ongoing. I will highlight some of the salient events in conference in which my experience of discovery began; I will describe my learning from these events—in particular, my experience of the intersection of the applied traditions of psychoanalysis and organizational role theory; and finally, I will present a case to highlight the dilemmas in role encountered at that intersection.

LEARNING TO LEARN FROM CONFERENCE EXPERIENCE

On the advice of friends and colleagues in the A. K. Rice Institute, I joined a membership of a 10-day residential conference at Yale University in the

---

1. The development of role theory and the evolution of the practice of role consultation have been informed by many, but prominent among them are Armstrong (1979), Lawrence (1979), and Reed (1976). Briefly defined, *role consultation* is a thinking partnership in which reflection and analysis of a role matrix (system in which role occurs, matrix of related roles, role stakeholders, role attributes, role requirements and expected behaviors, and skills readiness of the role occupant) are undertaken (1) in order for the role occupant to develop an alignment among overall organizational mission, expected role function, and person-in-role needs, and (2) to deepen an understanding of the overall system in which the role is imbedded.

fall of 1973. At the time, I had some passing interest in learning yet another "group technique," but I had no particular interest in, nor awareness of, the potential relevance that the conference might have to my personal development and to the developments in the various roles which I occupied in my organizational life. Nor, at that time, did I consider organizational life to be part of "real" life in any event. What went on at work was split off from what I considered to be my life; the "me" at work was more or less relegated to the domain of "not me." Organizational life was what one endured day after day, for a longer or shorter stretch, of oppressive containment. That my attitude, itself, might have interacted dynamically with and have had some collusive bearing on the conditions of poor practice, low morale, and deficient mission sense in my organization at that time, was wholly out of my awareness and experience. My "Tavi" friends had hoped that my attitude and unconsciousness could be penetrated by new experience in conference learning.

At the Yale conference, I was a member of what seemed like a pulsing mob of about seventy. We were mostly African-American members, ranging widely on what we then thought of as a "blackness continuum." Confronted with the task of working with a predominantly white-male staff, we enacted the social polemics and identity struggles of the day. Whether to join with white authority for mutual learning or whether to abort that process for safety-seeking guised as taking the opportunity to explore autonomous, independent, and separate black intellectual development became the terms of conference membership. It was during the conference-as-a-whole events where one experienced the anguish inherent in these terms. During the large group, one felt the tremendous internal pressure to join the hostile, silent, and consciously threatening majority that wished to exercise power for power's sake. While the attempt to erode authority and abort the task by refusal to join was a potent membership stance that was exhilarating, one felt anguish at the consequence of the destruction of thought and reflection. During the intergroup event, one felt similar internal pressure to join the hostile majority in abandoning the conference territory to caucus elsewhere in special interest groups where discussions of politically relevant issues of the day supplanted the assigned task of reflection on intergroup processes. We members intended to demonstrate the capacity to design learning opportunities specific to our racial interests, free from the control of authority deemed irrelevant by reason of its seemingly arcane conference mission. The cost of the feeling of belonging to one's identity group was that one experienced the anguish of the potential destruction of relatedness to staff and task.

The fear of the destruction of real and cherished relationships was no less tortuous, as the members of staff were my dear friends and close colleagues. During the institutional event, parts of the hostile majority set up an "alternative" conference staff to do the business of refereeing contentious debates on deeply held and widely varying political points of view about the workings of the possible solutions to institutional racism. While one experienced catharsis at bringing deep emotion and critical thinking together in the relative safety of the black-members-only configuration, one felt anguish about the constriction of learning, for neither staff nor members were at a point in cultural development where we could explore the political relatedness between the two groups and among the various subgroups that so suffused that conference.

The anguish experienced around the consequences of one's choices became the crucible for learning. The threats of violence and accusation of violence that occurred during the conference ignited my awareness that historical black injury and fears of competitive competence were pitted against longing for satisfying interdependency with white authority. The threats of "abortion of competence" and accusations of "abortion of competence" that occurred ignited my awareness that historical female injury and fears of loss of feminine "grace" were pitted against longing for successful collaboration with male authority. There was a giant man of giant intellect[2] at that conference as well, who along with other black members ignited the awareness of my dependency on and hatred of white authority that had long been covered by the defensive adaptation of white-identification.

In order to make sense out of the intensity of the culture of that temporary institution we were creating, I spoke out loud about my most intimate and primitive thoughts to strangers; I spoke of the love and hatred and longing I felt towards those authorities in suits and ties and high-heeled shoes; I blamed them for our condition of terror and helplessness that we were experiencing in the conference, and I wanted them to save us. In an attempt to escape the awfulness of the culture we were creating, I went to extreme lengths, right to the point of violating tenets of reality, to avoid assuming any accountability for the state of disrepair in the temporary

---

2. Leroy Wells, Fellow of the A. K. Rice Institute, passed away on Monday, October 20, 1997. I wish to acknowledge my gratitude for our chance meeting at the Yale conference. He, more than anyone, contributed to the foundation of my experiential learning in the life of the conference through his unrelenting intellectual and personal challenge coupled with unstinting moral support.

organization, and went equally as far to avoid taking any responsibility for its repair. My resistance to the immediacy of experience came in the form of rational questioning: "Weren't these ideas about 'black self' in relation to 'other,' ideas about 'me' and 'not me' simply my private intrapsychic dilemmas and private demons to be sorted out in psychoanalysis?" That these self-aspects might have analogues in others' experience, that they would have relevance in how I took up my roles, how I exercised my authority, and that they would interact dynamically with the efficacy of my organization, were revolutionary thoughts which had been elusively close to consciousness, and were so obvious once spoken, that the experience made me slightly heady. The exercise of putting these, and other of my inner experiences into the public domain for consideration in a collective attempt to understand living events (and, in this conference, to understand the common predicament) was affirming, authorizing, and exhilarating. It was the beginning of a critical understanding for me, of how I fit into the larger scheme. It was the first slip of the knot in the tangled warp of stereotypic self-assumptions about victimized and alienated racial specialness.

From the whorl of stunning ideas, high emotions, irrational feelings, and self-serving and anxiety-reducing assumptions on both sides of the task boundary came interpretations from the authorities and from the authorizing body which put a frame on experience, gave a language to the primitive, and informed the collective awareness regarding the link between the system we had formed and inner worlds that characterized us. It was very compelling. The interpretations, having the effect of incantations, lifted a veil between awareness and nonawareness, ignited an aspect of my intelligence, and moved my spirit (my childhood ambitions to be a Jesuit priest spoke of a readiness and capacity to be moved in my spirit by superordinate systems). It was the beginning of discovery, an experience of learning how to learn: linking my inner experience with the inner experience of others to develop new learning and shared understanding.

The conference as a whole articulated the system learning during the concluding review sessions. From the collective experiences of the destruction of thought and reflection, the potential destruction of relatedness, and the constriction of learning, we concluded that despite the catharsis of the '60s or because of the political disappointments then, there abided a deeply destructive racial split which none of the groupings—staff nor members, blacks nor whites—had the courage or the skills or the will to explore. It was a sobering conclusion, relevant to political and intergroup life in those times.

The implications of individual learning were developed in the role-analysis phase of the conference, where members were asked to reflect on behaviors, thoughts, and feelings experienced in the roles in which they found themselves during the conference and to link these to their organizational roles "back home." At that time, I was an assistant director of a higher education opportunities program, an affirmative action effort in the arts and science college at a local university. It was an underfunded program, floundering from the effects of a poorly understood and not well-articulated mission, and its culture was notably dispirited, primarily due to the high failure rate of its students. Key to my learning was the observation about the system in which I had designed to keep myself de-authorized: I had chosen a profession and an occupation because of its political relevance, but to which I had made uneasy and only partial commitments, as these choices were unsuited to my nascent wishes to shape, to manage, to join in creating, and to influence my working environment. I understood through the role analysis that I had placed myself in mediocre working circumstance with little authority and little opportunity to operationalize my own competence and value system and the value system that the program purported to represent, in order to keep alive a transference longing, and a vain hope that my talents would be discovered, a vain hope that I would be elevated, chosen, found, brought along, and given a place. The anguish at the disappointments in this regard supported my organizational transference, and served to confirm my stereotyped hypothesis that being both female and black constituted a double dose of containment and limitation in living and working. The idea that I might be maintaining a collusive posture vis-à-vis the stereotype as a security operation came as a rude (and a very welcome) awakening. Once again, it was a fact so obvious as to make me laugh out loud with my member cohorts in the role analysis in bittersweet self-recognition.

The role analysis consultant was instrumental in the discovery of the transference links to my role dissatisfactions and, unwittingly, I think, set the challenge for me to close the gap between discovery and action. He delivered a throwaway line that made a difference; he said: "Well, after a first conference, people do sometimes find themselves taking greater risks in being the architects of their own experience." Pathways had been laid down between my intense emotionality and primitivity and my capacity to think about my own and others' experiences in the world (of work), and to understand that my work was me, and that it did reflect the state of my relatedness to myself and to my surroundings, and that developing (this aspect of myself) was entirely in my hands and was limited only by the extent of my knowledge and awareness—which I could improve.

At this moment of new discovery, I was an analysand in the termination phase of an eight-year, five-times-a-week, classical Freudian analysis. The roles that I had taken up, the difficulties I was having with my own and others' authority, the ease with which I would let myself be defeated in attempts to move on task, the collusion to undermine authority through passivity and compliance with the status quo of dependency satisfaction, and the hatred I felt towards both my organization and its mission—none of these realities and bits of organizational transference had come to light in 2,000 hours of inquiry, but had become abundantly clear in a mere 10 days of conference work. When I took the learning about the metaphor of my organizational role dilemmas and its relationship to my early family roles back to the analysis, even my proper Freudian analyst got excited; he put aside interpretations about resistance and took my leadership in focusing on these bits of learning during the termination phase of our work.

As my "Tavi" friends had hoped, my armor of unawareness had been penetrated; my personal revolution had begun. I resigned from my job, started a small sustaining private practice of psychotherapy, and began my doctoral studies in preparation for analytic training. Given the new learning, I wanted not just any analytic training, but training in Interpersonal Psychoanalysis, where the training itself would afford me the opportunity to continue to deepen my understanding of the link between people and their systems. By its basic philosophy, Interpersonal Psychoanalysis authorizes a broader-than-orthodox range of inquiry into experiences in the extrapsychic world of role relatedness or powerful metaphors and analogue of early relationships, unresolved dynamic conflicts, and traumatic antecedents to personality dilemmas. Work roles and work relationships, I had learned in my conference experience, are as rich a place for inquiry as is the world of dreams and transference distortions.

CONFERENCE LEARNING APPLIED: THE LINK BETWEEN THE PRACTICE AREAS OF PSYCHOANALYSIS AND ORGANIZATIONAL ROLE ANALYSIS

The initial learning from the Yale conference has been the thread through my professional development. In particular, I discovered that the role analysis frame used in the conference was robust enough to allow for deep reflections in members, the kind of reflection usually thought to be the domain of the psychoanalytic frame; I discovered that this was a powerful frame for thought which provided a container for the preparation for psychic shifts and role shifts. I discovered in a very personal way that the learning from psychoanalysis and the learning from conference life—culminating in role analysis—were indistinguishable, and that the learning did not have to reside in the domain of private personal recovery—the learning could be applied. I discovered that the connections between self and system could be articulated and that reflection from either side of the continuum that connected self and system could inform understanding of either and/or both. I discovered that learning from each could be applied reciprocally. This series of discoveries, having led to many years of professional development, has culminated in my taking up a consultant role that contains both the practice areas of psychoanalysis and organizational role analysis or organizational role consultation.

Over the years, I have been working at the psychic integration of two practice areas, and I have found these to be mutually informing in useful ways. Rooted as I am in learning from experience and applying learning from experience, however, I am aware of a hesitancy to make explicit the learning that comes from psychoanalytic practice (and test out the possibilities) in reflecting on the practice of organizational role consultation. When ideas spring to mind that are generated from the accumulated experience of working as a psychoanalyst that might apply directly in role consultation or to consultant-team reflections, I experience anxiety. The fear is that ideas, if expressed, could potentially violate what feels like a taboo (not unlike the fears during my Yale conference). It is as if the learning might be so deeply personal as to have little applicability beyond observations on personal development. The knowledge that theory and competence are advanced from what is learned in practice abandons me. The converse is not true; learning from the practice of organizational consultation has profoundly affected my understanding of psychoanalytic practice— psychoanalytic contracting; locus of authority in psychoanalysis; and the learning partnership are some examples. I find these ideas relatively simple to articulate in professional psychoanalytic settings.

While I am curious as to whether the experience of constraint is self-generated and idiosyncratic or whether it is generated in some unconscious aspect of our culture of organizational consultation, I want to explore the constraint, and try to articulate the cross-learning from the practice areas. To begin, I would like to describe the similarities, differences, and overlap in the two practice areas of psychoanalysis and role consultation as I experience them, and present a case from which I would like to draw some inferences about the learning transfer between the two.

SIMILARITIES BETWEEN THE TWO PRACTICE AREAS

In terms of similarities between the two practice areas, there are two that seem obvious:

1. **Use of Self.** One uses oneself rather like a tuning fork in both practices—a place of resonance against which data takes on meaning that can be tested and refined and applied to the problem at hand. While the locus of psychoanalytic inquiry may be "tell me about your mother," and in role consultation, it is "tell me about your authority," one is available to understanding both the roots of the problem as well as its current manifestations in both practices.

2. **Work Occurs in a Dyad.** Psychoanalytic and role consultations are fundamentally undertaken in a dyadic task relationship[3] where utility is measured by successfully developed relatedness in the two-person system. However, the dyadic contexts or backgrounds are different in role consultation and analytic consultation. In role consultation, one holds vivid the organizational context in which the role occurs, and since authorization to do one's work derives from outside the dyad, either tacitly or explicitly, one may wish to and need to engage with those who people the context. In psychoanalysis, one holds vivid the familial organization in which the person evolved. While one engages the person's context intellectually or countertransferentially with a range of feelings, the actual people do not enter the sphere of the analytic consultation.

DIFFERENCES BETWEEN THE PRACTICE AREAS

Concerning the differences between the practice areas, there are several to be considered:

---

3. Role consultation can also be a useful diagnostic or intervention tool in groups, teams, and temporary conference institutions, where one works with the role occupant and focuses the relevant interactions and interrelatedness of the cluster of relevant roles to make system inferences, including one's role in the group.

1. Task Boundary. What differentiates the two processes of experiential learning are the task boundaries—the psychoanalytic task being recovery (where desire for recovery is driven by personal imperatives), and the role consultation task being a shift in role (where desire or requirement is driven by organizational imperatives). The differences in the task boundaries seem quite clear, although results may be indistinguishable: recovery in psychoanalysis often leads to shifts in work roles; and the outcome of role consultation can be the benefit of recovered well-being in the client in the context of organizational developments.

2. Authorization. By reason of task, the nature of authorization of the consultant differs as well in the two practices. Where the psychoanalyst is free to think and to say whatever deepens reflection (guided, of course, by constraints of the professional frame), the role consultant is constrained by contract focus and different terms of social propriety (e.g., where it would be quite appropriate in the psychoanalytic role, it might be a breach of propriety to inquire into sexual fantasies of a role consultation client, for example, even though these might be quite at the heart of stated role difficulties and account for some aspects of surrounding organizational chaos). In one recent role consultation, a diagnostic team came to understand that heightened sexual fantasies and destructive enactment by one member of the client group were the psychic strategies used to counteract experiences of vulnerability at feeling underskilled in his executive role. In a process of "role profiling"[4] that led to an acknowledgment of some of his real deficits, he could build-in supports for skill development. The role work produced a libidinal shift without direct inquiry into or interpretation of his psychological dynamics. With libidinal energy shifted from a narcissistic preoccupation to the task itself, the high-risk situation and chaos were ameliorated in his situation. The executive felt better as did his people and strategic developments in his sector of the organization could begin.

3. Confidentiality. Another difference in the practice areas concerns the nature of the boundaries of confidentiality. While confidentiality is a time-honored requirement in the dyadic task system in role consultation and psychoanalysis, the boundaries of confidentiality differ somewhat in each practice area. In psychoanalysis, confidentiality is expected to remain inviolate, except where there is explicit permission given by the patient in

---

4. Role profiling is a system of mapping tasks, responsibility, accountability, and skill requirements associated with a given organizational role.

extraordinary circumstances. Information derived, as well as assumptions and hypotheses made, are applied in such a way as to develop a shared reality between the analyst and patient in pursuit of recovery. Where role consultation is conducted in the context of larger organizational intervention, one's client is likely to be the manager of all the role-holders in role consultation. The principles learned in the role work might be shared with the client for the benefit of a larger system diagnosis in advance of developing an intervention. In this circumstance, the role consultation is a complementary task to the understanding of larger system dynamics, and confidentiality can be extended only in a limited way. While the particulars of an individual's role and person-in-role dilemmas may be held confidential, principles about the organization's culture, authorization structure, and so forth, derived from an understanding of dilemmas may not. The role consultant has to use critical judgment as to what advances the diagnosis and what is unnecessary divulgence. Initial contracting for this "almost confidential" work is critical and likely to be different, case by case.

4. Action. There is a fourth area in which the practice areas differ, and that regards "action." The role consultant is free to employ any suitable methods at hand, design interventions to help the client up the learning curve to "shift." The psychoanalyst is constrained against taking direct "action" on the theory that preservation of the opportunity to interpret irrational enactment in the transference/countertransference matrix is the tool of choice to help the patient learn. This constraint against direct intervention for preservation of the transference can become a defensive foxhole for the analyst if it is taken to mean that the inquiry must remain rarefied and exclude penetrating inquiry into the patient's real world of day-to-day experience in order to concentrate solely on the real world of internal fantasy, dreams, drives, and feelings.

In role consultation there is an expected outcome in accordance with a negotiated design with the client or client system. In psychoanalysis, there is no design for the shape of recovery; it is part of the wonder of psychoanalysis—that there are endless designs of discovery in the human spirit to emerge from the work.

AREAS OF OVERLAP

There are two areas of overlap in the practice areas that make role effects in psychoanalysis and role consultation indistinguishable:

1. **Idealization.** In both roles, the consultant has to work against idealization and attributions of "expert" where these are defensive. One would expect at least a minimum level of mutual "hopefulness" to be present as a necessary condition for successful contracting to give and receive service; however, "irrational hope" concerning the "expertise" of the role consultant or analyst can be problematic, signaling as it might, a passivity, an offloading of responsibility for change processes. We hear it often when patients want a "cure for my psychology," and clients want a "fix for my organization." In both practices, one takes up a bit of the teacher role to set the conditions for collaboration and learning partnership. Of course, one is able to do this bit of teaching if one is not struggling with "irrational certainty" in oneself. Whether irrational hope in the client or patient "induces" irrational certainty in either practice role, or the other way round, there is surely an expected interaction here, which can often have an unwanted consequence.

In the middle of the first phase of consultation, when we are working on the contract, and/or preparing for the first intervention, we can be stunned to find the client disappointed (or, in some cases, furious) about unmet expectations; one can find oneself having fallen a great distance in client's esteem (and the de-idealization can be a rather rude awakening). While a shift in valence is painful in any event, it may be more so for the lack of expectation of it. It seems to me that misalignment of expectations is an inevitable part of developing relatedness; as the relatedness develops and task-understanding deepens, latent intentions become manifest, or conscious intentions sharpen or expand. It seems to me that the question is not "will a shift in valence occur" but rather "when and how will it" occur.

In my experience in role consultation, the shift in valence occurs when one goes beyond the dyad and into the larger system; in psychoanalysis it occurs when the analyst is demystified and patient takes over the management of their own recovery. Of course, one has to work against grandiosity during periods of idealization in both practices, and one has to regulate the self-esteem when it changes—hopefully changes—in either role.

2. **Deep Knowing.** The second area of overlap in psychoanalysis and role consultation practice concerns the depth and source of knowing "the other." Both the role consultant and the psychoanalyst can come to know the client or patient at penetrating depths. Both can see the character or personality available to organizational dynamics or family system dynamics. In my opinion, the skill base for this capacity is quite the same. The role

consultant and analyst use an understanding of human psychology—derived wisdom from experience of personal, adult-developmental struggles—to connect to client and patient dilemmas. The difference in practice between psychoanalysis and role consultation concerns what one does with what one knows. In role consultation, deep and specific understanding of the client's dynamics are held to the background, used to inform discussion, design, and contracting efforts. A role consultant would not say to a client: "Your self-absorption may be creating problems with how your team is forming." Rather, a role consultant would take the client's particular idiosyncrasies into account in consulting processes (and laugh at his jokes). On the other hand, the psychoanalyst is contracted to develop deep knowledge and to use it explicitly.

THE CASE OF KARYN

With these similarities and differences in the two practice areas of experiential learning in mind, I would like to turn to the case of Karyn and Karyn and Company. I am presenting this case for two reasons: to consider the psychological roots of role dilemmas encountered by a woman in authority by telling the story of her analysis, and the story of her learning and my learning in it; and to reflect on dilemmas of the intersection of psychoanalytic and organizational role boundaries as these were confounded in our work together.

The Conscious Story

Karyn, a white woman in her fifties—tall slim, tailored, lovely—came for a consultation for analysis. She was tired, listless, burning out, feeling irritable and mildly depressed; she was unclear about her career goals. In the midst of a midlife reevaluation, she was somewhat alarmed to find an inclination to drop out and grow tulips. Karyn was then the co-owner and corporate president of a thriving enterprise in the communications industry. She is the co-founder of her organization, and since its beginning, 14 years ago, she has been working steadily—the long hours, late nights and weekends—on its development. She considered herself powerful, a driver, striving and ambitious, and competitively quite successful. However, her success was not as satisfying as it once had been. She remembered the days when she would get deep satisfaction as she measured herself against male rivals, former superiors and peers; or as she said: "Every drop of black ink shows the bastards." Her competitive strivings, once sustaining, now felt hollow. "Something was missing."

Also, during the consultation, she reviewed the history of her 20-year marriage to George, reevaluating their joint decision to forgo parenthood. Despite the stresses of two business careers, the couple had developed an emotional partnership and were good friends to each other; they were connected to their extended families, had an active social life, and kept up with their community and religious commitments. The marriage seemed as though it should be satisfying; it was difficult for her to account for the experience of conjugal emptiness. With regard to their childlessness, Karyn described having discovered in her first analysis the difficulty she had in identifying with her mother's deadened subjugation to the needs of a large family of children, and to the needs of a dominating, tyrannical, though quite dependent husband. As early as doll-playing years, Karyn had preferred to play "store" (her father owned a clothing store), buying, trading, bargaining, and stashing away pennies. She became father's favorite "son" and she ran the business for him for a time before her marriage and career.

Karyn had no model for managing others' dependency needs without loss of the sense of self. (Her first analyst gave birth to three children during the course of the analysis, but apparently, that model had not been sufficient!) She had decided in that analysis (and similarly, George, in his) that she was not inclined to childbearing nor to child rearing. The couple determined that they would create alternative means to satisfy their generative strivings; they would work creatively. (Her husband owns a business in a related field, and until recently, the couple spent a great deal of their time together brainstorming and problem-solving each other's business problems.) The decision has not been trouble-free. From time to time Karyn and her husband reconsider the idea of adopting an older child; they have sought consultation about this on two occasions. However, the conflicts about their childlessness have been external, presumably arising out of pressures from friends and family. The pressures notwithstanding, the couple continued to affirm their original decision. While there always has been a feeling of sadness around the edges of her adult life, Karyn has felt that her life choices, including this one, reflect a true sense of herself, both her limitations and her capacities. Nonetheless, "something is missing."

Reflections on the Conscious Story

The consultation led me to think that this was a midlife reevaluation in a male-identified woman who was having late-stage, generative strivings that were no longer being satisfied by her work. I made the assumption that the "something missing" was a deadened self-aspect. I assumed that she had developed a workable psychic consolidation, where she dissociated the experience of emptiness and loss through intense work-related activity, commitments, and over-functioning. As her energy waned, supports for the dissociative mechanisms eroded and troublesome feelings were coming conscious. I assumed that there were other life dynamics and psychic dynamics at work that were driving the erosion of defenses.

I admired Karyn for the acknowledgment of a developmental halt, and for wishing to restart her developmental engine at this stage in her life. I felt identified with her; I liked her very much and felt energized by the prospect of our working together. We began the analysis four times a week. Because she was experienced in the role of patient, we established a traditional frame easily. She worked with her dreams, linking current experience to past experience and acquainting me with her internal world and historical imagoes. Comfortably in pursuit of an understanding of what was "missing," I was intrigued by developing a picture of her relationships with parents and siblings, her early sexual fantasies, her obsessional preoccupations, and so forth. I became a benign object in the transference, it seemed—our relationship was most like the one she had with her paternal grandfather, who did not speak English, but with whom she spent many easy hours being baby-sat.

The Pseudo-Analysis

Given this experience of who I was to her in the transference, it shouldn't have done so, but it came as a surprise to me when she said towards the end of the first year that she was bored by our work, bored by the re-visiting of her history and the retelling of her fantasies. She felt that she was doing with me what she does in life—figuring out what "works" and doing that perfectly. We laughed in recognition when I made the observation that I had been unaware of her boredom, perhaps, because she was being a "perfect" picture of myself being a good patient. This countertransference insight led me to understand that we were engaging in a "pseudoanalysis"; that, likely, we were enacting the "something missing."

The Beginning of Work

I asked her: "If we were not being 'perfect' in our roles of 'good patient' and 'good analyst,' what would we be talking about?" "When we talk about my partner," she said, "we wouldn't take it to be a window to my past, we would talk strategy! We would be using these tools to help me think about the future. What should I do about my work? Why am I burning out? What would I do with myself if I retire? What alternatives are there that I can't see? George can't help me—like everybody else, he thinks I'm perfectly happy with my competent life. Margaret (her partner) can't help me—my confusion would threaten her future. If we weren't being so good, we would be helping me to think!"

A Frame for Thinking about Women's Development

Now, I am not unaccustomed to mentoring women patients in their work roles during an analysis. In my professional lifetime, the expansion of women's power and authority in the world of work has been explosive. I have given myself permission to work directly with women's articulation, adaptation, and psychic integration of the new learning required of them by the increased opportunities for professional growth and development.

The learning has been on several fronts: development of personal authority beyond traditional roles; self-esteem shifts to include conscious awareness of feelings of aggression, competition, and envy (and tolerance for others' competition and envy); learning to enjoy ambition, success, and money; learning to be at risk and accountable; learning to recognize and to develop strategies for using transference responses to their gender; and to manage their sexuality in role without managing away their femininity. However, there is one area that seems to defy easy adaptation to changing opportunities for some women. It is difficult to find syntonic solutions to disturbances in the relational field that the changed opportunities have wrought.[5]

Carol Gilligan (1993) provides a frame for this problem area that I find compelling. She puts forth an hypothesis that says that women's development is driven by the imperatives of caregiving and nurturing in the relational field, and these imperatives are at the center of women's self-identity. She continues by emphasizing that adolescence precipitates a relational crisis for girls. That is, a self-crisis for girls, where there is a necessary struggle to resist the psychological temptation to disconnect from feelings in order to preserve and to protect relations from the intense experience of love, lust, hatred, envy, and longing from the ordinary stuff of relations that girls are socialized to experience as antirelational.

---

5. In order to compete successfully for the opportunity to penetrate the relatively few cracks in the glass ceiling, a woman may have to compromise some aspects of nurturing and caregiving in her relationships.

Where girls are unable to resist the disconnection from feelings for the sake of relational preservation, there is a psychological wounding and concomitant suffering from loss of self-aspects—loss of authentic feelings; loss of connection to the body; and sometimes, loss of reality, but certainly, loss of the very meaning of relationship. Further, where girls at adolescence are unable to resist the disconnection from feelings, they may be pressed to take on images of perfection as the model of pure or perfectly good woman; the woman whom everyone will promote and value and want to be with.

The loss of self-aspects and the disconnection from true relatedness as a strategy to preserve the relational field may be a dynamic that becomes exacerbated in adolescence in women, but I believe the roots of relational crises in girls and women may be imbedded in earlier experience, as we will see as Karyn's case unfolds.

The Analyst's Resistance

So, here was Karyn describing disconnection from major relationships and libidinal withdrawal from her work, asking for concrete help in these areas in which I had developed a frame for thinking and had developed some experience in my professional role and personal life. Unaccountably, the opportunity to work with Karyn in this way filled me with dread. Regressive, defensive thoughts sprang to mind. I felt that I was in the wrong role to do strategy. On the one hand, because of my organizational background and the nature of her industry, I felt engaged by the prospect of doing strategic thinking with her. But on the other, what would this have to do with psychoanalysis, and with the "something missing"? My internal good standing as a psychoanalyst also felt threatened no matter which direction I took. If I expanded the inquiry to think directly, not analytically, about the dilemmas before her, I would be acting out. If I made standard resistance interpretations, I would go against my deeply held value that the patient is often right, knowing intuitively what is needed for their psychic development.

I considered taking my dread and anxiety for consultation. I could hear Dr. Favorite Senior saying, "She invites you to shore up a schizoid consolidation. She wishes to defeat the analysis. Interpret her resistance." Or, "You have a narcissistic, grandiose wish to be all things to her. You wish to defeat the analysis. Consider your resistance; perhaps you're bored with psychoanalysis altogether."

Anxious about this continuum of possibilities, I did not go for consultation; rather, I contemplated the obvious—perhaps I should refer her for business consultation, or I should terminate the analysis and do the consultation myself. I equivocated privately about this dilemma for a couple of months. I overcame my sense of dread with an internal reassurance that while I was helping her to think about the reality of her situation, I would be developing a picture of who she was and how she functioned. We both had a highly developed analytic frame-in-the-mind with which to hold onto the primary task of recovery of the "something missing." My need to develop these rationalizations, the lack of fluidity in expanding the boundary of the inquiry, and my experience of trepidation became catalogued as data.

Shift in the Inquiry

I shifted the inquiry from elaboration of early-childhood fantasy to a closer focus on the operations of her company. I found that the organization had two structures—a formal one and an informal one.

Formal Structure. In the formal structure the organization provided support services: public relations and media relations for small communications organizations, especially where there were mergers, acquisitions, and divestitures. The company consisted of two partners, six senior staff, eleven associates, and an administrative staff of seven—all women. There was a long-term clientele and a growing list of new clients—predominately male. Since the partners had come up through the ranks in the communications world, the company enjoyed a good reputation on a wide network. The company was also financially solid, although they did have a rough period during the 1980s when the communications industry experienced near catastrophe.

Informal Structure. The informal structure of the company, on the other hand, was quite different. I found that Karyn and Margaret, because of their success as entrepreneurs in their industry, were much sought after by other women colleagues who wanted their advice, coaching, and direct support in their careers and in business enterprises. The partners responded to this interest in them with great vitality; they generated sophisticated and creative strategic interventions in this work with colleagues. They consulted to women's issues of accretion of power in their institutions. They held small group seminars on the topic. They developed strategies for financial recovery of one woman's small business; they helped her find the right consulting firm by interviewing potential consultants with her. The partners strategized the maintenance of her corporate credibility, when another woman intended to balance her staff in favor of more women than men. They coached their women on personality traits—temper, passivity, appeasement tendencies, timidity, or overaggressiveness; they advised them as well as on "dressing for power" (it was not unusual for them to shop with a woman friend to help her buy clothes that could make the right statement at a critical presentation). They also coached one woman to launch a sex-discrimination suit—

the first in this woman's industry—and remained her shadow consultants during the many months of high-profile, stressful proceedings.

Their women were free to call on them any time, business hours or otherwise. They were in great demand, and often heard after some success "I couldn't have done it without you." Despite the pleasures and rewards of gratitude, Karyn and Margaret were doing this informal work at the expense of personal time and time for reflection. Along with the concerns of their own business, they were swamped. Karyn, in particular, had recreated her mother's sense of drowning in others' dependency; in her case, it was others' work dependency. Her primary work task was being drained of energy and vitality, and Karyn simply wanted to withdraw. Given this understanding of her situation and psychology, I made the direct suggestion in the analysis that Karyn and her partner consider what were the impediments to expanding their business to include a consultation service; that is, what were the impediments to legitimating the function to which they were naturally inclined, for which they had some talent, and for which they had a clientele. "You mean make it real?" Karyn thought this was an idea worth pursuing and asked me if I could talk with the partners as they elaborated this potential.

Dread returned and my anxiety intensified. I think my suggestion to diversify the business had an instruction imbedded in it: "Would you please fix this, and let's get back to traditional psychoanalysis!" I think I dreaded what I might discover. In order to think this through with them, I would have to hear the stories of the many women in a less derivative way and I thought I might find the stories distasteful—stories of women who were feeling depleted, feeling inauthentic; women who were idealizing and frightened of male authority; women who were seeking approval and hurting. I thought it would be hard to bear. In the company of one woman at a time attempting to create or invent herself—as is the case in psychoanalysis—an identification is in place that makes the pain of visiting these experiences in myself tolerable. In the company of many women in the struggle of self-creation, I worried that I would be swamped with feelings of sadness for myself and sadness for them that it is such a struggle. I was worried that I might find this aspect of their enterprise distasteful, and I wanted to protect Karyn and myself from this possibility.

The Change in the Analytic Frame

My worries notwithstanding, we terminated the analytic work. We reset the frame: I wrote a letter of intent, outlining timing and content of an assessment, re-did the fee structure, and changed the billing from Karyn to the company. I went to their place of business to do the assessment. I worked with the partners as a pair to elaborate the impediments, as they saw them, and to work out the meaning of a potential expansion of the business. Some of the issues that emerged for them were concerned with the political meaning of shifting from sisterly sharing on the old girl's network to charging fees for service, and the fear that they might lose credibility and be seen as base, unfeeling, and opportunistic. As well as a necessary differentiation in the partners' roles, and an expansion of personnel, also there would be the additional work of codifying their experience of doing role consultation intuitively in order to market it. They would have to make it real, and were somewhat worried about an insufficient skill base with which to proceed. They also worked on the question of whether they would have a market beyond their immediate collegial network and their referrals. Who else might need or want the service? Who else was providing the service, formally or informally, and how would they figure this out? Who was their competition? (It had not occurred to me until this writing that I was her competition!) What help would they need?

This part of the work with the partners was rather easy and straightforward elucidation of strategy. However, when I observed several of their coaching sessions to understand firsthand the intuitive methods of role consultation they were using, my worries were realized; I found Karyn's style of coaching extremely distasteful. She was vituperative, attacking, and assaultive.

For example, a corporate vice-president for public relations came for advice because she was suffering an erosion of credibility. She had "unwittingly" become the confidant of the members of her executive team, including the chief financial officer, confounding her work task by intervening in interpersonal difficulties. She was fatigued and frightened that she had been indiscreet in matters of confidentiality, particularly of her own. She realized that in a current political skirmish, she was going to lose control of her budget. She wanted to resign before she suffered the indignity; she felt that she could not bear it. She felt humiliated, that she had been so naive in how she took up her role. In response to the story, Karyn says "Ach! Women. Can't take it on the chin! You stay in there. We'll figure out how you get your budget back." In assessing her client's situation, Karyn told her that in her anxiety to gain acceptability on the team, she had used her womanly wiles—providing comfort and empathy—and had gotten herself converted into some version of mother or wife or sister by her colleagues, and had become an easy target for competitive gambits in a predominantly male setting. She had hoped to gain a greater position of influence, status, and power by making herself emotionally indispensable to the team. "This is a woman's gambit," Karyn told her. "Stupid! Not only did you make them envious, you lost your effectiveness. This 'empathic readiness' B.S. is dangerous." Karyn suggested that the woman hire a consultant for her executive team, and back out of being "Ms. Fix It." "Do your caregiving at home!" Karyn told her.

While I thought Karyn's observations about the woman's role-dilemmas were insightful and the recommendation well-placed, I found the abusiveness of her client's style, and Karyn's contempt for her feelings of vulnerability intolerable.

As part of the assessment, I had interviews with her clientele to get their picture of the service. I asked about the harshness and abusiveness. Basically, what they were aware of was the relief to have penetrating attention to the detail of their real-life development issues. "That's just Karyn. She hasn't been wrong where it counts." As this was the end of the assessment phase of work with the partners, I suggested that if they were to diversify the business to include consultation services, they might take their own counsel and hire someone to help them elaborate and refine their intervention strategies. I suggested some training possibilities, gave them Carol Gilligan to read, and referred them to a team of three consultants, two women and one man; the team contained role consultation, leadership development, and technical capabilities among them.

I wrote a working note framing my concerns about the efficacy of her abusiveness with future clients, concerns for that part of her market that did not include colleagues and acquaintances already conversant with her style. I suggested several courses of action to meet the partners' need to develop a skill base in their consulting style. I sent my bill and was paid.

Return to the Analytic Frame

Karyn and I returned to the four-times-a-week analysis, and I began to work with her on the abuse of the "womanly wiles." I told her that I noticed how angry she seemed to be at her clients' "woman stuff." I told her that I had a dream during the assessment, and that it had brought up a memory for me: When I was about 4 years old, I told my father that I didn't want to be a mommy, that I was going to be a priest when I grew up. He was a little alarmed. He told me to wait and see how I felt when I was a little older, but he did not think that girls could be priests. "Why not? I don't want to be an 'ucky' mommy!" "That's not my little girl," he said, and asked me to hand him his newspaper. I had told my dearest parent, my pal, my daddy, that there was something deeply disturbing about being a woman-to-be, and he had wanted no part of this part of me. And so it was gone forever. In the company of women, I had recovered this relationship to my woman-self that had long been forgotten, that had not been recovered in two analyses—neither with a

male analyst, nor a female analyst. I had steered through my various life pas-
sages with something disconnected.

I asked her if she had such an experience. Karyn recalled a story that
her mother told her—that Karyn had fainted at the sight of an undressed man-
nequin when she was "real young" (mother did not ever say the age). She
said, "I don't remember this, but I do remember a bad feeling—like being
very sick to the stomach when I saw my pregnant mother nude" (being the
oldest of six, she had many of these opportunities). She felt that her mother's
body was grotesque—all that hair and fat and flab. She dreaded the day when
she would have those things on her chest, heavy and horrible. She hated the
whole idea; she hated her mother for being that way; she hated her mother
for making her a girl. She remembers hearing her parents having sex and
hearing her mother laughing and thought, "she's just pretending." She thought
of her mother like a marionette, being jerked through life according to someone else's will—having babies, having sex, being agreeable, being dependent,
being mindless of all the greedy need for her. She hated her vulnerability,
hated her need to make a cold and distant husband feel indispensable. She
hated the untruth of being a woman.

Karyn was incredulous at the discrepancy between what these memories revealed of her inner state and the conscious picture she held of herself.
She was incredulous to find hateful feelings towards her mother; fears of feeling vulnerable and hatred of her own feelings of vulnerability; hatred of her
mother's body, disgust with her own body-to-be; hatred of her dependency
and longing for a reliable dependent relation—all masked and dissociated.
She understood the power of the psychic assault that had resulted in the
disconnection of herself from herself and in the fervor with which she tried
to stamp out these feelings and experiences in her women clients. She said,
"I didn't want them to feel so scared. I was trying to yell some strength into
them." Karyn had immersed herself in the company of women in her informal structure in search of her woman-self. The intensity of her immersion in
this conflict was the main drain on her energy and the source of the burnout.

The Real Analysis

This set of insights launched the real psychoanalysis. At this state, it became
clear to me what had occasioned my dread and anxiety in the countertransference. Karyn and I had a deep identification as women who had made an
early disconnection from feelings about womanness (in my case, to preserve
my relation to my father as an ideal little girl; and in her case, to preserve her
mother's ideal image of herself in Karyn, and therefore, their relationship).
Far from the pseudoanalysis, authentic work and connection with Karyn

would require a reconnection to disturbing feelings in myself and I would have to contend with disturbing feelings that would emerge in the transference. I felt unsure of myself in the absence of an experiential road map.

And indeed the transference was hateful and vilifying. This was an intense period in the work, where we each were feeling quite vulnerable. Being in relationship to her while she recovered these feelings awoke more of my own memories. I felt vulnerable to the precision of her insights about my limitations, and I felt sad about her losses and my own. Karyn had made herself vulnerable by risking the ruination of our relationship in feeling and expressing feelings of hatred towards me. I think that tolerating the unaccustomed feelings of vulnerability between us was key in Karyn's work of recovery. As our relationship did not deteriorate, her long-ago developed hypothesis that she had to be out of herself to maintain the ties eroded. She developed an understanding of the strategy that distanced her from hateful feelings and led her to develop the persona of a strong, competent, reliable, courageous person on whom one could rely, but whom no one would ever know. To maintain her strategy, she had to forgo deep intimacy, feelings of vulnerability, and, certainly she could not have risked the necessarily vulnerable times of childbirth and child rearing, as the unwanted feelings might swamp her consciousness.

These self-insights produced a protracted period of mourning. The period of mourning and reintegration lasted for approximately two-and-a-half years. When we terminated the analysis, Karyn and I were both more conversant with the "something missing"—it was a connection to troublesome feelings about being a woman.

THE ONGOING LEARNING

Not unlike the experiential discoveries at the Yale Conference, I have learned a great deal from my experience of working with Karyn and Karyn and Company. While there are ambiguities to be resolved about role contamination, loss of distance, unconscious countertransference effects, and dependency gratification that led me to take on the dual role of analyst and role consultant in this case, the experience has sharpened my thinking about some aspects of the two practice areas:

1. There Is Transfer of Learning from Psychoanalytic Practice to Role Consultation Practice

In her work *Women, Girls, and Psychotherapy: Reframing Resistance*, Carol Gilligan (1991) says that women who struggle with resisting the psychological temptation to disconnect from feelings as a strategy to preserve the relational field, often find themselves in therapy for one form or another for emotional, political, or organizational trouble. Or, I would add, find themselves in a role consultation.

We are all familiar with the idea that power striving in some women may be an antidote for feeling empty and disconnected, where they can use potency and competence as defense, a reaction against feelings of vulnerability, something akin to identification with the aggressor. At one time, I would have described these women, myself included, as male-identified women, that is, women who had primary identifications with idealized or troublesome paternal attributes. I would refine that, now, to say that the dynamic is better described as maternal, counter-identification, where women develop traits of aloofness, distance, and disconnection (not unlike stereotypic male attributions) in pursuit of relational preservation. This distinction is essential when working with women in authority, as I think is expressed in the following brief vignette:

> A woman executive in the financial-services industry hired me to work with her senior team. In a period of explosive growth, she was hiring managers who, once on board, became tentative and dependent on her explicit directions. She found this intolerable, as she wanted them to form into a team and get going.

Senior team members thought of her as distant and impossible, and basically wanted to stay out of her way. With an eye on assisting with team formation, I found myself coaching new hires around her "bossy, aggressive personality" on the theory that they were experiencing frustrated dependency strivings towards a male-identified woman in the context of an intensely evolving task system. I was wrong. It occurred to me that she was attempting to establish and to maintain relatedness with her people by using role consultation like an adapter pin. I would be the communications link and her surrogate in task relatedness. I took from my experience with Karyn the awareness that I needed to refocus the role consultation to determine what relational preservation strategies might be alive in this situation. I gave her the hypothesis about dilemmas in the relational field that women may experience and asked her to discuss with me what impediments she saw in working directly with each of her seniors and with them as a group. This set of discussions was useful; it became clear that she was protecting them and herself and their future together by not exposing her feelings of dependency on them at this critical juncture in the company history. I continued to work with this client by consulting to the executive team, which did form and which included her. (The job was less about providing the relatedness link, but naming this relational problem and helping her develop risk-diminished strategies to work with the problem.)

With this experience as a guide, I am guided to add to my repertoire of reflections when I respond to a request for role consultation with a woman in authority, not only with the question: "How does this woman manage her authority?" but also "Where is she in this likely common dilemma of disconnection for preservation in the relational field?"

2. The Two Practice Areas Are Mutually Informative

The confounding of psychoanalysis and role consultation in my work with Karyn has drawn a sharp picture for me about the ways in which the two practices are mutually informative. I think that I abandoned the analytic frame or confounded the roles of analyst and role consultant in this case because I held the analytic role inflexibly. While I hardly subscribe to the current and popular notion that psychoanalysis is dead or dying, I had held the analytic role in a deadened way. In the role consultation, I was able to perceive—fully—some aspects of Karyn that had not been available to me in the analysis. I take this to be a comment on a self-imposed constraint against massaging the boundary of the analytic inquiry. Analytic curiosity about her clientele, I think, would have yielded the same understanding that I abandoned the frame to be free to pursue. When I ask myself the question about why I did not continue the role consultation once I had made an observation that might have been useful in the partners' designing process, I conclude that I had not authorized myself sufficiently as a role consultant to "know what I knew" about Karyn's abusiveness deeply. I had enough data to "know" from Karyn's vituperative projections onto her clients that she was attacking relatedness, and that she was operating from a female counter-identified place; I had my dream and memories to support that conscious observation. I could have coached-down her vituperativeness. With humor and common-sense language, I could have given her an hypothesis to consider about this difficult aspect of herself being expressed in the way she was relating to her clients. I could have generalized from my own memories the reasons why she might be having these difficulties.

(I certainly have done all this subsequently with women—and it does not take deep analysis for these conversations to produce insight and role shift.) I could have . . . well, I could have done a number of things, but the fact that I resumed the analysis is the third point of my learning.

3. Task Interdependence Is a Distinct Characteristic of the Practice of Psychoanalysis

There is no doubt that the task of Karyn's recovery had not yet concluded when we returned to the analysis. But a deeper truth is that I was also in the process of recovering lost aspects of self in the analytic task with her. And it was not concluded. In my quest to differentiate the psychoanalytic and organizational role consultant roles so that I can better integrate them, the experience of working with Karyn sharpens one of the critical differences between the two practice areas: A byproduct of the intensity of psychoanalysis is a task interdependence that develops between the patient and analyst: "I can help you recover to the extent that I am recovered; your recovery may awaken unrecovered self aspects in me while we are in the task relationship." This bit of contracting is unnecessary to the task of role consultation, and therefore it does not become, as psychoanalysis does, a "safe house" for the personal development of the analyst as well as the patient.

CONCLUSION

The purpose of this paper has been to trace the threads of my experiential learning in the Tavistock tradition, at the Yale Conference some 20 years ago, through to a reflection on the application of the learning transfer between the practice areas of psychoanalysis and organizational role analysis. Certainly, the model of learning that I discovered there, one that links inner experience and inner constructs of reality to processes of evolution, development, and change has been important to development of my two professional roles, and it has helped me begin to articulate their similarities, differences, and dilemmas. Additionally, and perhaps, more importantly to my personal development, the model of learning from experience has given me a way to keep my opportunity for discovery alive and the possibility of acquiring new learning ongoing.

REFERENCES

Armstrong, D. (1997). The institution in the mind. *Free Associations* 7:48.

Butts, H. (1971). Psychoanalysis and unconscious racism. *Journal of Contemporary Psychotherapy* 3:67–81.

Butts, H., and Schachter, M. (1968). Transference and countertransference in interracial analyses. *Journal of the American Psychoanalytic Association* 16:67–81.

De Geus, A. (1988). Planning as learning. *Harvard Business Review* 66(2):70–78.

Duhl, B. S. (1983). *From the Inside Out and Other Metaphors.* New York: Brunner/Mazel.

Gilligan, C. (1993). *In a Different Voice.* Cambridge, MA: Harvard University Press.

Gilligan, C., Lyons, N. P., and Hammer, T. J. (1990). *Making Connections: The Relational Worlds of Adolescent Girls at Emma Willard School.* Cambridge, MA: Harvard University Press.

Gilligan, C., Rogers, A., and Tolman, D. (1991). *Women, Girls, and Psychotherapy: Reframing Resistance.* Boston: Harrington Park Press.

Grinberg, L. (1979). Projective counteridentification and countertransference. In *Countertransference*, ed. L. Epstein and A. Feiner, pp. 169–193. New York: Jason Aronson.

Issacharoff, A. (1979). Barriers to knowing. In *Countertransference*, ed. L. Epstein and A. Feiner, pp. 27–45. New York: Jason Aronson.

Kohut, H. (1984). *How Does Analysis Cure?* A. Goldberg and P. Stepansky (Eds.). Chicago: University of Chicago Press.

Lawrence, W. G. (1979). A concept for today: the management of oneself in role. In *Exploring Individual and Organizational Boundaries*, ed. W. G. Lawrence, pp. 235–249. New York: John Wiley & Sons.

Reed, B. (1976). Organizational role analysis. In *Developing Skills in Managers*, ed. C. L. Cooper, pp. 89–102. London: The Macmillan Press.

Witenberg, E. The inner experience of the psychoanalyst. In *Countertransference*, ed. L. Epstein and A. Feiner, pp. 45–59. New York: Jason Aronson.

2

The Exercise of Authority in a Dependent Context

WESLEY CARR

THE BACKGROUND TO DEPENDENCE

To be dependent is the natural human condition. No one chooses to be born and no one can be born without a mother.[1] In most eras to recognize such obvious dependence and its lifelong impact and psychological significance was normal and more personally and socially acceptable than is allowed in late twentieth century Western societies. With its emphasis on interrelationship and connections, the Enlightenment made dependence an issue. It was one among many topics in human behaviour that was categorized and explored. Now, in so-called postmodern times, the emphasis is shifting to a preoccupation with the self and its perceptions. Any connection with another person or object, therefore, becomes suspect. As a result dependence, which always involves connection, is confused with addiction and no difference is discerned between, for example, relying on a tradition, seeking the approval of others, or using alcohol or drugs. Even what is natural is assumed, often unconsciously, to be a malign state (Kegan 1995). The political use of such phrases as "the dependency culture" has enlarged the scope of the theme. Dependence has moved from being assumed to being analyzed and finally to being regarded as an undesirable facet of life. It is, therefore, widely believed,

---

1. This statement is becoming increasingly qualified by modern techniques of reproduction. But it remains true that whatever the technology involved in anyone's origins, he or she did not choose to be born.

implicitly and explicitly, that any shift from dependency to autonomy is both desirable and an achievement.

Yet there remain some obvious contexts in which dependence in its most simple form has to be mobilized. For instance, medical treatment involves putting oneself in the hands of someone else, whether trusting a diagnosis or, when being anaesthetized, physically surrendering consciousness. Doctors utilize this dependence in treatment, even though it may sometimes be abused. Acknowledging such dependence is not immature behaviour: critically to trust a doctor is evidence of maturity. Even more obvious is religion. Its power to legitimize inevitable human dependence was an early object of sociological study (Durkheim 1915). This aspect also fascinated Freud, Jung, and many of their successors (Freud 1927, 1930, Jung 1964, Rizzuto 1979, Badcock 1980, Meissner 1984).

How authority is exercised is also of particular interest in relation to dependence. Most obviously it may take the perverse form of authoritarianism when a leader colludes with the primitive dependent expectations of a group, institution, or society. In this chapter I shall seek to clarify the concept of dependence as a distinctive context for the exercise of authority. Two case studies are offered, one illustrating the complexity of dependence when authority is powerfully assigned and one exploring how in a profoundly dependent context (a church), the leaders were able to exercise authority by paying careful attention to the process and avoiding either collusion with or dismissal of dependent expectations.

BASIC ASSUMPTION DEPENDENCE

The tradition out of which this chapter is written is that associated with the Tavistock Institute of Human Relations. One major contribution to this synthesis is W. R. Bion's seminal thinking on the dynamics of groups (1961), and its being regularly re-explored in group relations conferences (Rice 1965, Miller 1989). Bion's clarification of the distinction between the work group and the basic assumption group is crucial (Bion 1961). "Work" describes that aspect of a group's life that is designed to achieve a task—what there is to do. In such a mode it may be called "a work group". A group's basic assumption life is its emotional existence, usually unconscious and unspoken. This may occasionally be harnessed to task performance, but more often seems to hinder it, especially if left unaddressed.

Of course, there are not two groups: rather the two modes of functioning are intertwined in the dynamic generated by one set of participants.

One of the three basic assumptions that Bion identified is that of dependence. In this mode the group regresses to seek its gratification by creating a leader, sometimes the designated one, sometimes not. For many, regression is associated with a childlike denial of responsibility. Bion quotes a member of such a group. Asked why he was not contributing, he replied, "I do not need to talk because I know that I only have to come here long enough and all my questions will be answered without my having to do anything" (Bion 1961, 147f).

A considerable number of those who have studied this field have a background in psychotherapy. They, therefore, almost instinctively tend to interpret such behaviour in the same way as they would that of an individual in a therapeutic setting. Not surprisingly, therefore, to be dependent is usually regarded as a state from which the person or group needs rescuing. Indeed, some characterize the state as malign:

> The "dependency" group perceives the leader as omnipotent and omniscient while considering themselves inadequate, immature and incompetent. . . . Thus primitive idealization, projected omnipotence, denial, envy and greed, together with defenses against these, characterize the dependency group, and its members feel united by a common sense of demand that it preferred not to address into those who for their own needfulness, helplessness, and fear of an outside world vaguely experienced as empty or frustrating. [Kernberg 1978]

I have often consulted to groups functioning in a dependent mode. But while such extreme behaviour is occasionally encountered, life in the group is usually more subtle. Care is even more important when we turn to application of these insights. For it is in connection with the social aspect of dependence that the issue of the exercise of authority arises. What, for instance, of a group the task of which (the function of its work group aspect) is specifically to manage some form or forms of dependence? Following and elaborating Freud (1921), Bion suggested that certain social institutions may be designed to deal structurally with basic assumption life. Freud had seen that the army and the church were particularly prone to responding to and stimulating activity respectively in the basic assumptions of fight/flight and dependence. Bion suggested that they might be "budded off" by the main group, of which they form a part, for the specific purpose of neutralizing dependent and fight-flight groups. Thus they may prevent these dynamics from obstructing the work-group function of the main group. But to do so they not only have to receive the projections from society at large, they also have to work on them. Since the internal and external worlds of institutions are dynamically complementary (Miller and Rice 1967, Shapiro and Carr 1991), he would expect to see dependent behaviour, for example, in a church as it handles dependency.

DEPENDENCE IN SOCIETY: THE JAPANESE EXAMPLE

Scholars have chiefly derived data on dependence from the Western psychoanalytic and sociological tradition. But the dynamic is not confined to those worlds. Takeo Doi (1973) has explored the Japanese concept of *amae*. His translator, John Bester, defines this as referring "initially to the feelings that all normal infants at the breast harbor towards the mother— dependence, the desire to be passively loved,[2] the unwillingness to be separated from the warm mother-child circle and cast into a world of objective 'reality'" (p. 7).

This study is especially instructive for two reasons. First, for Western readers it examines dependence from a perspective that is both foreign and familiar. The culture of Japanese society is different from that of those in the West. Therefore to discern an underlying psychological phenomenon in such diverse cultures invites comparison and contrast. Second, it draws attention to the importance of cultural variants when we consider the dynamics that underlie the life of any human group. This is a dimension in the study of groups and their dynamics that may sometimes be overlooked and its significance underestimated.[3]

Authority has to be acknowledged in Japan in such a dependent culture. Doi indicates that the role of the emperor makes this possible. He embodies the dependent/counterdependent relationship:

> The emperor is in a position to expect that those about him will attend to all matters great and small . . . In one sense he is entirely dependent on those about him, yet status-wise it is those about him who are sub-

---

2. The concept of "passive love" is derived from Michael Balint (1969).
3. Membership of the Leicester Conferences for the study of group relations has become noticeably multinational, although this form of study has not yet been replicated in Japan or China (Miller 1989).

ordinate to the emperor. . . . the person who can embody infantile dependence in its purest form is most qualified to stand at the top in Japanese society. [p. 58]

Although the emperor system as an ideology may be in decline, *amae*
remains central to society. What is in danger of being lost is not social
dependence but structural ways of enabling it to be addressed and positively
harnessed. Doi points out that the emperor's role is twofold: he both embodies societal dependence and enables it to be used in Japanese society.
One characteristic of such dependence is in the way that the group is
customarily assigned priority over the individual. This is specially interesting for contemporary Western society with its major emphasis on the
autonomy of the individual and the idea of personal authority. Yet at the
same time there is increasing anxiety about what society itself is and what
sort of society is desired.

A key issue amid the confusion over dependence is how authority
can be acknowledged in anyone. Doi does not suggest that the group is
"better" or more significant than the individual in a moral sense. But he
argues that enabled dependence always takes corporate form. He contrasts
Western and Japanese values. The emperor and the family system, before
the end of the Second World War, had together imposed ideological
restrictions on behaviour in society. The decline of these influences had
not, however, as is commonly assumed in Western contexts, "served the
cause of individualism, but by destroying the traditional channels of *amae*
had contributed, if anything, to the spiritual and social confusion"
(p. 21).

The life of the individual, therefore, whatever the context, involves a
basic dependence that will find some sort of expression in the group. The
term "group" can be enlarged to include "society". That this dependence is
acknowledged and managed is important not only for the welfare of individuals but also of organizations, institutions, and societies. At least it cannot
be ignored.

Without a recognized focal point (or set of points) embodying and
thus enabling the acknowledgment of dependence, Doi argues, people
regress to dependent nonwork states. When this happens a society operates
basic assumptions alone and authority is consequently a greater problem.
Leadership becomes a longed-for but incomprehensible ideal because
followership, without some accompanying appreciation of dependence, is
an impossible concept. Without resorting to apocalyptic language, we may

then ask what sort of social cohesion is possible, even if all agree that it is desirable.[4]

DISTINGUISHING DEPENDENCE FROM DEPENDENCY

A distinction may usefully be drawn between "dependence" and "dependency". These words are often used interchangeably. Bion in his seminal work wrote of the dependent basic assumption as "dependence".[5] Thirty years later Miller's (1993) illuminating collection of papers on application and consultation from Bion's conceptual perspective employs "dependency". But both are discussing the same phenomenon. To clarify this complex dynamic in what follows I wish to distinguish "dependency" from "dependence". From the perspective of any exercise of authority in a dependent context, this distinction is both useful and essential.[6]

Whatever autonomy we may each succeed in achieving, we remain interdependent with our environment. Exceptions to his claim might be suggested. Do not recluses, for example, renounce any interdependence and cut themselves off from the world? But when interpreted dynamically, the hermit's role seems to perform a function on behalf of the majority who are not solitary. Historically, recluses sustained a distinctive, if by today's standards curious, ideal of holiness. The majority in society projected aspects of religious demand that they preferred not to address into those who, for their own psychological reasons, were willing to bear them. But all interdependence implies dependence (Gould 1999). Dependent existence is neither good nor bad: it simply "is". This state we may call "dependence". It begins at birth and we oscillate in relation to it, from time to time regressing and then shifting towards autonomous positions. Such behaviour is not just a matter of survival. It is generative, since it is through working out these dependent relations to the world and to others that people acquire and discover their values.

---

4. For an example of how this question may be currently addressed, see the articles by Shapiro and by Obholzer in Shapiro (1997).

5. In fact, for the most part, he used the adjective 'dependent' and avoided the nouns in favour of his own technical shorthand (baD).

6. I am indebted for the distinction to discussions with Bruce Reed of The Grubb Institute and an unpublished paper of his (1995).

"Dependence" is a human characteristic, a property of both individuals and groups. In group relations conferences some conventions of everyday life are removed or ignored in order to enable the participants to explore and study unconscious processes. But even in that stressful and rigorous setting recognition has to be given to this dependence. The members of such conferences, for example, reasonably expect the conference staff to create and sustain certain boundary conditions on which they can rely. Without these the study (that is, the task of the conference) cannot occur. The ambient factors for work are maternal—a reliable schedule that is adhered to, suitable space for the work to take place, food, and accommodation. They constitute the holding environment that invites confident regression by the members in these aspects of their life so that they can face the discomfort of less familiar dimensions as these are exposed.[7]

"Dependence" is to be contrasted with "dependency". When people in a group become anxious about survival, they unconsciously employ various devices—the basic assumption behaviour that Bion discerned. One, possibly the most common, is to identify someone on whom to depend. He or she will usually have some presumed authority, such as, for example, that of the consultant to the group (usually in such mode unconsciously treated as its leader). They then behave as if this person were all powerful and they were unthinkingly dependent. Interpretation may bring them back to reality. But as Reed (1995) points out, "those involved are not usually changed by this experience and in their various settings in different groupings they continue to seek defensive stratagems against their anxieties".

Troubles ensue when people reject (or feel they are required to resist) dependence because they cannot distinguish it from such dependency. They then become locked into a cycle. Reed again remarks: "Rejecting dependence as an unacceptable relationship, they are caught up in the unconscious process of dependency and through the experience of their powerlessness hate it. . . . Wishing to be independent the only form of dependence they can experience is 'immature dependence'—or dependency".

Dependence and dependency are obviously connected. But it is important to distinguish them, and particularly to show which is being re-

---

7. The notion of a holding environment describes the management of emotional aspects of family life, which makes individual development possible. The concept was proposed and elaborated by D. W. Winnicott (1960). Further applications of the concept may be found in Shapiro & Carr (1991), especially Chapter 3, and Gould (1999).

ferred to when the adjective "dependent" is employed.[8] If it refers to the unconscious surrender of authority, then it can be addressed through interpretation and a shift from dependency to dependence and some sort of autonomy may be made. If, however, it describes a manifestation of appropriate dependence, then it is itself an exercise of authority. The following case illustrates the difference and the potential confusions.

CASE STUDY 1: A TEACHING SESSION

The director of a hospital with a noted reputation for treatment, training, and research invited me to consult to a group of trainee therapists. All were already qualified as psychiatrists or psychologists. They were completing their therapeutic training, which included a weekly session with the director. He was a distinguished practitioner, author, and teacher in his field. The trainees consequently looked up to him and were inclined to model themselves on his style, both consciously, and no doubt even more so, unconsciously. One persistent and common difficulty emerged that was not peculiar to these particular trainees: How were they simultaneously to manage themselves in two roles? In one they were trainees—people who were publicly learning to be therapists and being taught. But in order to be trainees they also had to take up the role of therapist. They were each assigned a patient in the hospital for whose treatment they, in collaboration with others and under supervision, were responsible. They experienced a conflict of authority between the two roles and could not see an obvious way of resolving it.

By contrast—and for the trainees discomfortingly—the director was notably clear on the issue of authority. It was, as the trainees and the staff generally perceived, "his" theme. This made him a useful and valued tutor, whom the trainees used as a benchmark for competence. They judged themselves and others by this standard. He was also, however, held in awe for his proficiency. As the students grappled with their dilemmas, they time and again spoke about having to "take your own authority". The phrase seemed to be used in a variety of settings. A distinguished visiting scholar, for example,

---

8. Sometimes the terms "immature dependence" and "mature dependence" are found in the literature. They roughly correspond to the distinction that I am drawing between "dependence" (mature) and "dependency" (immature).

had offered a different set of approaches to therapy, which had disturbed the trainees' familiar world and made them anxious. It was to be relieved by taking their own authority. The same was true of dealing with supervisors, colleagues, and even agreeing or disagreeing with the director. Any dilemma and they were to take their own authority. But the phrase had a specious ring to it. On the one hand it was "right": the issues concerned authority and the trainees had to discover how to exercise it in a variety of settings. On the other hand it seemed also to be used to discount the appropriate and mature dependence that characterizes collaboration between teachers and learners, supervisors and trainees, and director and staff.

After hearing this mantra repeated for a while, I asked the group where they got it from and what it meant. The phrase, of course, was the director's, who frequently offered it as advice. Initially the trainees seemed shocked at my question. They regarded it as a challenge to the director, which was itself evidence of a confusion between their dependence as learners and their dependency in the group. I persisted that I was not asking about the director's problem but theirs. The response eventually emerged: no one was sure what the command meant. All, however, reckoned that it was a lifeline in moments of confusion and that within it probably lay the secret to becoming both a good learner and a good therapist.

Case Study 1: Discussion

The instruction to take their authority was being used to contain and resolve the stress of the experience of conflicting roles. The director had intended it to be a way of shifting the trainees from their dependency on him and the structures of the hospital into appropriate dependence. For this they would require a developed sense of role. When, as was inevitable, role conflicts erupted, they would then be noticed and competently handled. But the phrase had become an incantation and was no longer (if it ever had been) a working idea. It was, therefore, being used to reinforce dependency. Instead of scrutinizing their roles, the students were trying to acquire the director's competence by assuming his language and consequently, as they unconsciously hoped, his power. They were using uninterpreted dependency as a way of avoiding the demands of dependence. As a result they had lost sight of their primary but necessarily conflicting roles of trainee and therapist. In these they had authority to learn and authority to treat, such authority deriving from structured and acknowledged dependence. But they were hearing the injunction in terms of their person and finding their dependency reinforced.

The Boundary between Dependence and Dependency

The boundary between dependence and dependency is usually subtle and needs monitoring. It is here that, as the case above suggests, the widely used contemporary phrase "personal authority" becomes risky. Miller conceptualizes "'personal authority' as a function of managing oneself in relation to role and task performance, while 'power' is concerned with maintenance and enhancement of status and with control over other people" (Miller 1993, p. 310). This comment occurs in a discussion of what he calls "the post-dependency society". It may be that this is dawning. Certainly there are major changes in the structure of work and leisure, as well as economics and politics. It is also true that "commitment to task is sorely needed by the managements of today's organizations. . . . obedience is the kiss of organizational death."[9] The question to be addressed, however, is what happens to the distinction between dependence and dependency when dependency is resisted or denied? We might further surmise that any attempt to shift directly from dependency to autonomy would be experienced as the denial of dependence. In that case we would hypothesize a growing fear of association or collaboration with others; a refusal to countenance interdependence; and ultimately the confusion of authority with power that leads to authoritarianism. Indeed Miller himself implicitly suggests this with his critique of the fashionable vogue for empowerment:

> The theme of helping people to gain greater influence over their environment underlies virtually all of my work in action research and consultancy as well as education and training. Nowadays some people would call this 'empowerment'. It is a term I avoid because of its ambiguity: between *becoming* more powerful and *making* more powerful. The notion of giving power is inherently patronizing—it implies dependency—and hence is of itself *disempowering*. . . . power and dependency are central issues for a consultant working with organizations. [Miller 1993, p. xvi]

---

9. The theologian Donald McKinnon at Cambridge University used to say, "obedience is the most deceitful of the virtues."

Authority in a Specifically Dependent Context—The Church

Authority is an aspect of a person's role and that may only be discovered in relation to a task. This observation in turn brings us back to institutions in society and their organization. In order, therefore, to explore how authority functions in the context of dependence and dependency it is useful to study an institution that operates with these dynamics. As we have noted, both Freud and Bion suggested that the task of the church is distinctively located in this area.

Their hypothesis that society uses the church to handle many of its dependent aspects matches these institutions' self-understanding. The arguments are complex and beyond the scope of this chapter. But they would be expressed in theological, historical, and sociological language. Briefly, the theology is that churches offer themselves to be used by others, and put themselves in the way of so doing, because they believe that this is congruent with the nature of God. Historically, Western societies have been predominantly Christian, and hence there is a long-standing expectation that churches will be there and available. Both these perspectives can also illuminate the motivation of churches and the way in which they may merely by their existence provoke guilt and encourage dependency. Sociologically, the discussion revolves around the question of why religion continues and the way that it takes institutional shape in churches. How these functions may now be shared with other institutions in a society is a matter of debate. But there remains evidence that one of the functions of churches in contemporary society, at least those that have not become sectarian, is still in the field of handling not just dependency but also the boundary between dependency and dependence.

This suggestion refines that of Freud, Bion, and Reed regarding the task of the church in socety. We may hypothesize that, as institutions, churches have a distinctive task of handling dependency and transforming it to dependence. They are also institutions in which the exercise of authority is both complex but also fairly accessible for scrutiny. And for a society, even one dominated by secular and pluralist assumptions, it can also be argued that management of that shift from dependency to dependence is essential to its well-being. Doi does this for Japan. I have argued the case for England elsewhere (Carr 1985a, 1992), as has Modood (1994, 1997), who has also argued strongly that competent performance of this task by the indigenous British churches, especially the established Church of England, is essential for the religious integration of the emerging plural society in Great Britain.

Studies of behaviour in and around religious institutions frequently focus on the psychological effect on individuals of their belief and affiliation. There is, however, another dimension of behaviour to be studied. This is the intergroup dimension, the way in which churches function in and interact with society.[10] For this exploration the primary focus is on groups and their functioning, and consequently upon the roles that individuals may take up or may be assigned. The largest and most amorphous of such groups is "society". But even this can be interpreted in dynamic terms (e.g., Reed 1978, Khaleelee and Miller 1985, Shapiro and Carr 1991, Carr 1993).

All churches and religious bodies, like institutions in general, exist in interaction with their context. But the nature of that interaction varies according to the historical, sociological, and psychological factors that apply. And for churches, as noted above, the way that they theologically construe their identity contributes to the negotiation of this interaction. In England, for historical, theological, and cultural reasons, the Church of England distinctively offers itself to work in the context of social manifestations of dependency. Insofar as people, whatever their status in society and their personal beliefs, still come to the church at significant moments in their lives and that of the nation, the authority of that church to minister in this fashion seems confirmed. This claim becomes clear when questions of membership are raised. People do belong to congregations. But the ministrations of the church, especially of its clergy, are not restricted to these alone. All people are regarded as having a claim on this church, unless they opt not to exercise it (Carr 1992).

The key organizational unit of the Church of England is the *parish*, a defined area that is usually geographical, although it may also be a human zone such as a school or a factory. The local *vicar* (priest or minister) is responsible for the spiritual well-being of the people who live and work there, even if they do not regularly worship. The church building is often a focal point for community life, not only church activities. Families that do not necessarily attend church from time to time look to the church and

---

10. The remark of Miller and Rice (1967, p. 17) is apposite: "The individual is the creature of the group, the group of the individual."

vicar for services, especially at birth (baptism) and death (funerals). These are known as "the occasional offices" (Carr 1985b).[11]

In this context authority is especially exercised by the minister or priest. Lawrence and Miller (1973) have provided a perceptive insight into the minister's role and his or her authority in a dependent environment. I have often used it with clergy and laypeople in various parishes and contexts and found that, in spite of social changes over the past 20 years, it still resonates with their experience. Discussing the requirement for preparing men (and subsequently, of course, women) for the public ministry of the church, the authors disregarded the question of personal qualities and directed their thinking to the boundary at which ministers in role and their dynamic environment interact. This they characterized as dependent: "The minister, as he goes about his job as a representative of the Church on the boundary with the rest of society, has a great deal of hope invested in him" (Miller 1993, p. 106). This investment is not made by believers or worshippers alone; it spreads through communities (in the broadest sense of that term) and manifests itself often unexpectedly (e.g., Reed 1978, Ecclestone 1988, Carr 1993).

In one study I spoke to a woman who vehemently denied that she had, or wished to have, anything to do with her local church and vicar. Yet when I suggested that the logic of her position was that I should recommend to the bishop that he should withdraw the priest and close the church, her immediate and instinctive response was that she did not mean that. It was important for the town to have the church presence and to "know that there was someone praying for us [sic]". This story illustrates Lawrence and Miller's further contention: "There is inevitably an element of childlike dependency in the relationship to the Church, and thus to its representatives, in that to some extent they are being asked to solve the insoluble, cure the incurable, make reality go away. . . . Ministers of the Church, then, have to receive this dependency" (p. 106).

This is an apt (note especially "childlike dependency") description of the context in which churches work and in which authority has to be utilized. The effectiveness of this exercise varies: it was to clarify this area and better prepare people for ministry in it that Lawrence's and Miller's consultation was requested. "Sometimes they [ministers] get stuck in a paternalistic posture; sometimes they are able to help their parishioners both to

---

11. Twenty-seven percent of babies born alive are baptized in the Church of England (1996); many funerals are conducted by the clergy, although fewer marriages. Other notable occasions are times of major disaster or celebration. See Angela Tilby in Carr (1992, p. 83).

recognize their dependency and to discover their own resources and capabilities" (p. 106). How to exercise authority in a setting so suffused with dependency is illustrated in the following case study.

CASE STUDY 2: A PARISH CHURCH

The parish concerned was large, consisting of about thirty thousand people. Many of these were young, newly married, or in partnerships. The birthrate was correspondingly high. The population was mostly working class, many of them having their roots in East London, which was about 35 miles away. Families had been relocated after the Second World War and a second wave of migrations had occurred about 10 years before the study. The community, therefore, seemed to wish both to replicate life as it was believed to have been in London—to do the right thing—and to establish their own new and distinctive style of life. It was caught in an oscillation between dependency (preserving the past) and autonomy (developing the new life).

Church attendance was not high. Yet about one thousand people a year—between fifteen and thirty each week—were approaching the clergy to enquire about the christening (baptism) of their babies.[12] The clergy were expected to respond warmly and helpfully, and generally did. Indeed they could have spent all their time at this work. The applicants were in a deep mode of dependency. Usually it was the mother who came, and the characteristic dependent behaviour was manifest: acute sensitivity to any sign of rejection or affronting behaviour; self-justification; claims that someone else was behind the request—often the mother-in-law, and so on. All of these behavioural characteristics are familiar to family therapists (e.g., Box et al. 1981). The applicants were coming from their mostly God-ignoring world with a sense that something ought to be done which was "proper" or "right". They, therefore, approached the God-person (the vicar in the mind), whom they were expecting to act for them:

> People seeking baptism arrive with a complex set of notions. . . . Under-
> lying all, however, is the generalized idea of "God" or "the church". It is
> a mark of human experience that such massive notions are likely to be
> most effectively addressed if they can be in a sense located. Since they

---

12. "Christening" is the popular word for the Christian rite of initiation for babies; "baptism" is the technical ecclesiastical word for the same rite.

represent an awesome dimension to life, with which in the nature of the case the applicants only rarely become involved, part of the belief system is that authorized ministers are those who are best qualified to manage them. It is a mark of dependence that people construct in their minds a hierarchy of competence and authority. . . . The minister as "God person" . . . is a key component in ministering with many people, including those who bring their children for baptism. A publicly authorized minister, therefore, is needed in the approach phase. [Carr 1985b, p. 73]

A dynamic interpretation of the church's work is that it is to allow people to express their dependency and to interpret it into dependence. These clergy knew both instinctively and as a result of their training that this was one of their roles. They also were aware that their authority for such activity was firm. They had been ordained by the church to this ministry, licensed by the bishop to this place, and confirmed by the fact that people approached them to perform it. But while willing, indeed eager, to work with dependence, they could not cope with the demand, both physical and emotional, of such overwhelming expression of dependency. It seemed to require their presence: no one else would do than a "proper vicar". This demand was presented in an imperious fashion that was both aggressive but also passive.[13] They equally knew that they could not risk appearing to affront such tentative approaches and simply resisting the applicants' dependency. That would alienate people from the church. And given the number of people involved in the community, they might even potentially generate, or at least contribute to, social rage. With consultation, an elegant solution emerged.

The vicar advertised that each Wednesday the church would be open to deal with christenings. The clergy guaranteed the presence of at least one of them—there would therefore be an authorized minister—a vicar— available. He or she would be assisted by members of the congregation. People considering christening were encouraged to bring anyone they wished—parents, friends, and especially the baby and any siblings. Responsibility was thus already being laid on the family as a whole, however fragmented, and not just on one parent, usually the mother. This was the beginning of a process of shifting people from dependency to acknowledge

---

13. For further exploration of this phenomenon see Ecclestone (1988), especially the chapters by A. W. Carr ("Working with Dependency and Keeping Sane") and J. V. Taylor ("Conversion to the World").

their dependence. All were initially welcomed by the lay members of the congregation. Care was taken that a good number of these were themselves parents of young children and so knew about life in the community from that angle. The clergy on duty maintained a high profile: they wore clerical dress and moved around the hall, avoiding being stuck with any one person or group. Meanwhile in a side room trained lay people met the applicants and took basic details—names, addresses, and so forth.

One condition for any baptism was that applicants should participate in a preparation programme. Most people were willing to accept this. The requirements were outlined at this initial meeting. Most importantly, however, the applicants were also given a date for the baptism. Assuming that the programme was completed, applicants could be sure that the rite would be celebrated. Thus basic anxieties were met and the parents could become responsible for organizing their life towards a specific date for the ceremony. In terms of process their dependency was beginning to be turned to dependence, a form of collaborative interdependent working with the church.

Towards the end of the evening, at a previously advertised time for both starting and finishing, everyone was invited into the church. There the vicar led a brief, simple service. This followed a standard, formal structure as a thanksgiving for the birth of a child. Invariably people, even those who were struggling with the demand for preparation, expressed appreciation of this. During the following few days one of the clergy visited the home, over the weeks parents took part in the church programme, and eventually the christening took place. For a few this brief service offered what they were looking for and baptism did not ensue.

Case Study 2: Discussion

This case illustrates how through the exercise of authority in a highly dependent context people may shift from dependency to dependence and acquire the corresponding ability to take responsibility for their lives (that is, exercise their own authority). But because the connection with the church and what was expected of it concerned dependence, the notion of autonomy is here inappropriate. Responsible behaviour meant being able to collaborate in handling acknowledged dependence.

Where in this scenario did authority lie and how was it exercised? The issue was, as always, complicated. The vicar was publicly authorized with the church's authority to baptize. But this authority had to be recognized and affirmed by the applicants. In so doing they subtly affected that authority. For them in their dependency the vicar was looked to as the God-person. Such authority borders on power and its nature is such that it feels as though it can scarcely be scrutinized. It is, therefore, crucial that this person, who is the focus for so many projections, is self-conscious in role and does not relapse into simple use of the person. The parents, too, possessed their authority and on its basis they were seeking something for their child. For them too, role was central. By approaching the church, even in this highly dependent fashion they were tentatively beginning a new exploration of their authority in role as parents. They were discovering in particular how they might act on behalf of their child in an area of life—the spiritual or religious—that made them nervous because of their believed incompetence. Their behaviour, therefore, although they might not have realized it, was authoritative insofar as they could work in role. Indeed, for them personal belief, for instance, could not be the basis for engagement with the church. Peripherally there were the other church members and the group association of those also seeking baptism. The way that authority was exercised in this dependent environment took three phases.

First, dependency was recognized and affirmed. The clergy acknowledged a primitive and unexamined reliance on them as representative figures in the awesome area of God, church, religion, and the meaning of new life—the birth of the child. Into that mélange it would have been difficult, probably impossible, to inject immediate interpretation. Therefore, they created a holding environment of a published day, a reliable presence, and a structure, particularly a firm ending time. They thus met the prevailing dependency by being available and not being themselves disturbed about it. Their sense of role was accordingly high, so that they could act with authority without colluding with the applicants' dependency. But by making that availability identifiable and reliable they began the process of responding. On that basis they were able to engage with people, to interpret the dominant dynamic and thus to bring about a desired transformation. People working in dependent contexts often seem to confuse availability and reliability: dependency seems to demand the former, but what is being sought is the latter. It is an aspect of that early dependent posture towards parents, which has to be learned then and relearned at points in life.

Second, the interpretative skill was to shift that dependency into a usable form of dependence. An interaction between the context and the parents was needed that would enable them to begin to attempt their own interpretation of why they had come: that is, their exercise of authority in role was crucial to their shifting from dependency to dependence. This was achieved through two invitations. The first was to the preparation programme, which required the parents on their own authority to engage with the church. But such responsibility cannot be loaded on people when it is dynamically difficult for all, and probably impossible for most, to take it up. Hence a structure was created that offered another reliable point around which to make that decision. This was the promised date for the proposed ceremony. Here we see dependency, the debilitating surrender of responsibility, being transformed into dependence, a mature recognition of reality.

The second invitation was the immediate offer of the thanksgiving service. It provided an interpretative moment and a working space. In it people began to recognize that they were making their own decisions and interpretations about their approach without requiring final decision. Here again dependency was being transformed into mature dependence, which in this context acknowledged that work would be achieved only by practising interdependence. It was a means by which the authority of parents was both confirmed and worked on as it engaged with the sought authority of the church. They thus became appropriately dependent: that is, they were acknowledged as responsible for their decisions, but in a context where such decisions were beyond them. It would be wrong to call this autonomy. For these people were necessarily in a dependent mode. But a shift from immature behaviour (dependency) to a mature sense of responsibility within the given context (dependence) marked this point. The applicants needed, and would always need, the church for the task in hand—the christening. But the dependent and counterdependent relationship between applicants and church could be maturely affirmed by all involved in it.

The third phase was when the twin authorities—that of the church (through the minister) and that of the parents—came together to generate the baptism service itself. By that time leadership and followership were being exercised by both parties. The church was both leading by performing the ritual and following by responding to the parents' informed request; the parents were leading by taking limited responsibility for their decision and for their child and following by conforming to the programme of preparation. It would, therefore, also be false to describe this as an expression of autonomy. In a dependent context autonomy is unlikely. There is rather a negotiated interpretation not just of what needs to be done but of the dynamic environment itself.[14]

---

14. See "On 'Negotiated Interpretation'", Shapiro and Carr (1991).

CONCLUSION

Somewhere Hannah Arendt has remarked that the loss of authority is the loss of the groundwork of the world. Authority is itself a negotiated concept. One reason why it is frequently confused with power is because this aspect is overlooked. Because it is negotiated, the dynamic context in which it is being exercised is a contributory factor both to the nature of authority and the ways it can be exercised. This area, which is frequently left unexamined, is one that is exposed for study in the Tavistock-style group relations conferences. The brochures that advertise these conferences usually contain some words to the following effect:

> We may scrutinize overt political processes, in which, for example, ideologies are advanced to justify either maintaining or changing the distribution of power. But if we are to do this, we need also to examine the implicit assumptions and myths that reinforce the ways in which we relate to one another as individuals and groups. Consciously determined policies are sometimes supported and often subverted by these less conscious factors. *Whether we are in the position of leadership or followership, a responsible exercise of authority requires us as individuals to become aware of our own contribution to these conscious and unconscious processes.*[15]

In, for example, military leadership, the exercise of authority is often close to that of power. Orders are given and are in most circumstances expected to be obeyed. Although soldiers are today encouraged to think for themselves, they do so within a strict regime of clear relationships between leaders and followers. Mostly this is carried through, particularly in battle; otherwise death may result. By contrast, the exercise of authority in an institution such as a school, hospital, or church is more often marked by negotiation.

But in the light of the foregoing discussion it is not surprising that all such institutions can be, and sometimes are, marked by authoritarianism. One reason is that those in them, and those looking to them, confuse their natural dependence with dependency. Then this may be abused by the leader who operates on his or her personal power or charisma, rather than through interpretation in role. Because of the personal gratification to the

---

15. The author, Anton Obholzer (1997), was the Conference Director. [My italics.]

one exercising authority and the psychological relief, albeit temporary, to those vesting it in him or her, there is always a near danger of confusing the exercise of authority with authoritarianism. That this matters for society as well as for its own sake with individuals and institutions is perhaps now obvious. To be wary of dependency should not lead to fear of dependence. As a negotiated concept, authority is especially crucial in such settings. Failure confuses people, who, whatever they may eventually say, look to the authority figure to act on and with the authority that they have assigned to him or her.

Anyone who is working with dependency and dependence and who sensitively claims or is assigned authority must behave in ways that invite scrutiny and correction. But even more important, they must invite further exploration of the issues that involve all—teacher and student, priest and parishioner, consultant and client. That is how authority may be confidently exercised in the context of dependency, and even in the face of dependence.

REFERENCES

Badcock, C. R. (1980). *The Psychoanalysis of Culture.* Oxford: Blackwell.

Balint, M. (1969). *Primary Love and Psychoanalytic Technique.* New York: Liveright.

Bion, W. R. (1961). *Experiences in Groups and Other Papers.* London: Tavistock Publications.

Box, S., Copley, B., Magagna, J., and Moustaki, E. (1981). *Psychotherapy with Families: An Analytic Approach.* London: Routledge and Kegan Paul.

Bruce, S. (1995). *Religion in Modern Britain.* Oxford: Oxford University Press.

Carr, A. W. (1985a). *The Priestlike Task.* London: SPCK.

———— (1985b). *Brief Encounters: Pastoral Ministry through Baptisms, Weddings and Funerals.* London: SPCK. Revised edition, 1994.

———— (1987). Irrationality in religion. In *Irrationality in Social and Organizational Life*, ed. J. Krantz. Washington: A. K. Rice Institute. A revised version appears in A. W. Carr (1997), *A Handbook of Pastoral Studies.* London: SPCK.

———— (1992). *Say One for Me.* London: SPCK.

———— (1993). Some consequences of conceiving society as a large group. *Group: The Journal of the Eastern Group Psychotherapy Society* 17:235–244.

Davie, G. (1994). *Religion in Britain Since 1945: Believing Without Belonging.* Oxford: Blackwell.

Doi, T. (1973). *The Anatomy of Dependence.* Tokyo and New York: Kodansha International.

Durkheim, E (1915). *The Elementary Forms of the Religious Life*. London: George Allen and Unwin.

Ecclestone, G., ed. (1988). *The Parish Church?* London: Mowbrays.

Freud, S. (1921). Group psychology and the analysis of the ego. *Standard Edition* 18:69–146.

———— (1927). The future of an illusion. *Standard Edition* 21:1.

———— (1930). Civilization and its discontents. *Standard Edition* 21:59.

Gould, L. J., Ebers, R., and Clinchy, R. (1999). The systems psychodynamics of a joint venture: anxiety, social defenses and the management of mutual dependence. *Human Relations* 52(6):697–722.

Jung, C. G. (1964). *Man and his Symbols*. New York: Dell.

Kegan, R. (1995). *In Over Our Heads: The Mental Demands of Modern Life*. Cambridge and London: Harvard University Press.

Kernberg, O. (1978). Leadership and organized functioning: organizational regression. *International Journal of Group Psychotherapy* 11:3–25, cited in M. Pines, ed. *Bion and Group Psychotherapy*, pp. 140–143. London: Routledge and Kegan Paul.

Khaleelee, O., and Miller, E. J. (1985). Beyond the small group: society as an intelligible field of study. In *Bion and Group Psychotherapy*, ed. M. Pines, pp. 354–385. London: Routledge and Kegan Paul.

Lawrence, W. G., and Miller, E. J. (1973). The diocese of Chelmsford: a preliminary study of the organization for education and training in the context of the task of ministry. In *From Dependency to Autonomy: Studies in Organization and Change*. London: Free Association Press, 1993, pp. 102–119.

Meissner, W. W. (1984). *Psychoanalysis and Religious Experience*. New Haven, CT: Yale University Press.

Miller, E. J. (1989). The "Leicester" model: experiential study of group and organizational processes. *Occasional Paper No. 10*. London: The Tavistock Institute of Human Relations.

Miller, E. J., and Rice, A. K. (1967). *Systems of Organization: The Control of Task and Sentient Boundaries*. London: Tavistock Press.

Modood, T. (1994). Ethno-religious minorities, secularism and the British state. *The British Political Quarterly* 65:53–65.

————, ed. (1997). *Church, State and Religious Minorities*. London: Policy Studies Institute.

Obholzer, A. (1997). *Authority, Leadership and Organization: A Working Conference*. London: The Tavistock Institute Group Relations Programme and The Tavistock Clinic Foundation.

Reed, B. D. (1978). *The Dynamics of Religion. Process and Movement in Christian Churches*. London: Darton, Longman and Todd.

———— (1995). *"Dependence" and "dependency"*. Unpublished paper.

Rice, A. K. (1965). *Learning for Leadership: Interpersonal and Intergroup Relations*. London: Tavistock Publications.

Rizzuto, A.-M. (1979). *The Birth of the Living God.* Chicago: University of Chicago Press.

Shapiro, E. R., ed. (1997). *The Inner World in the Outer World: Psychoanalytic Perspectives.* New Haven and London: Yale University Press.

Shapiro, E. R., and Carr, A. W. (1991). *Lost in Familiar Places: Creating New Connections between the Individual and Society.* New Haven and London: Yale University Press.

Winnicott, D. W. (1960). The theory of the parent–infant relationship. *International Journal of Psycho-Analysis* 41:585–623.

3

An Organization Looks at Itself: Psychoanalytic and Group Relations Perspectives on Facilitating Organizational Transition

RINA BAR-LEV ELIELI

INTRODUCTION

The need to stop for reflection and review with the purpose of gaining better understanding is not often recognized by organizations in the process of transition. Being in a process of transition, in itself, creates an atmosphere of moving rather than of holding or stopping. The notion of movement, transition, change, and shift is not compatible with the notion of stopping, taking a break, reflecting, and reviewing, even though whenever we think about "transition" and "change" the idea of resistance to change comes to mind.

We could say that in the "unconscious mind" of the organization, in the minds of those whose role it is to facilitate action in order to make things happen, transitions are related mainly to outside reality, while reflecting and reviewing is related mainly to internal reality. It is hard to conceive that both facilitating action and reflecting can relate to the external and the internal reality at the same time. It is often difficult to remember that institutions of all kinds leave their marks on the internal as well as external reality of individuals and organizations alike.

Whereas the "doing" aspects are more obvious since they are in line with the declared goals and tasks of the organization, the "being" aspects are more covert, and are frequently conceived as being disturbing because dealing with them may not be in line with the pursuit of productivity. This can be said even about organizations whose main productivity lies within the realm of human care, like health, education, or welfare, where the results of work are often intangible and difficult to understand.

This chapter will tell the story of an organization that was facing a major transitional change and did, indeed, take a break for one day to look at itself, reflect, and review. The story will be told from the point of view of the consultant who was called to intervene in what was felt by the management to be a major crisis in the midst of the transitional process. It will focus on the day of the review as well as on the consultants' meetings with management before and after. The review became the starting point of an ongoing process of consultation on different levels of the organization.

Before telling the story, as I perceived it, let me say a few words about some aspects, relating to the role of the consultant. W. R. Bion (1977) wrote:

> When the analyst is not sure what it is obtruding he is in the position of having an intuition without any corresponding concept—that intuition might be called "blind." Any concept, for example projective identification, is empty when it has no content. The problem for the practicing analyst is how to match his hunch, or his intuition, or his suspicion, with some formulations, some conceptual statement. He has to do it before he can give an interpretation. The analyst's role, in other words, is one which inevitably involves the use of transitive ideas or ideas in transit. The analysand, likewise, is attempting through his free associations to formulate an experience of which he is aware. [*Caesura*, pp. 43–44]

I could very easily replace the words "analyst" and "analysand" with the words "consultant" and "client," to describe my state of mind when taking up the role of a consultant. The fact that the role of the consultant implies its transitional aspects, as Bion calls them "transitive ideas" or "ideas in transit," defines for me the domain in which one meets the client and the process of consultancy even when called in for a consultation that does not relate to transition in the usual sense of the word. By a certain way of listening, the consultant can lead his client, an individual or a group, into a state of self-attention and self-awareness. This is how the client will be able to make sense of his own experiences, emotional and others, in order to regain his sense of destination, rather than be attached only to the "tragic position" of his state of being (Lawrence 1997), to the feeling of inevitable doom, often emanating from unconscious self-destruction.

THE CLIENT

The client, called a "station" by its staff, is a clinic, belonging to a network of clinics that provides psychological, psychotherapeutic, consultative, and organizational services to a large number of boarding schools and day schools for children and adolescents. These children and adolescents come from families of immigrants and/or broken families whose children had to be placed outside their home for various mental, economic, or legal reasons. Some of the children were sent to Israel while their families were left in the countries of origin, for example, the former Soviet Union, the former Yugoslavia, and Ethiopia. The network was created by, and belonged to, The Jewish Agency for Palestine, an international Jewish organization which has dealt with Jewish settlement in Palestine since 1929.

The staff is composed of thirty social workers, clinical psychologists, and interns. The management is composed of three people: The director, a male clinical psychologist; the chief clinical psychologist; and the chief social worker, both females. The chief social worker was a newcomer to the management. She replaced a staff member who had recently left because of the transition.

Almost a year before I was called in, it was announced that the whole network would be handed over by the Jewish Agency for Palestine to the Ministry of Education and Culture. The decision on this step was taken by Prime Minister Izhak Rabin the night before he was assassinated. People say it was the last decision he made. As Rabin's assassination was a major national event, this timing shed horror on the transition. It was clearly a nationwide decision, one that touched the very core of the life of the young state. The network had been crucial in rescuing and bringing children to Israel during and after the Holocaust. The context of the decision related to the changing relationship between Israel and the Jewish Diaspora, a highly emotional and politically loaded issue. The link between these two entities, Israel and the Jewish Diaspora, was no longer what it used to be.

The director of the station had contacted me immediately after the announcement was made. At that time I proposed to them that we work through the transitional issues on the level of the whole national network. For many reasons this did not work out and the consultation was carried out with only one node of the network.

FIRST MEETING WITH THE MANAGEMENT

I met the management once before commencing the intervention with the whole unit. In this meeting I could sense a feeling of depression and loss. The management did not know how to lead their staff. They surely did not wish to lead them anywhere. They could not identify with the political decision that the organization be handed over to the Ministry of Education and Culture. They had no sense of direction or destination. They hardly met representatives of their new parent organization, let alone the person who would be responsible for managing them. They had no regular management meetings. All their discussions amongst themselves, as well as with the whole staff, revealed no differentiation in role definitions, status, or hierarchy. All decisions were taken in a "democratic" way. The unwritten rule of this organization was never to offend or hurt anyone. It felt like meeting the grieving helpless parents of a boundaryless family who lost some of its best members, mainly old ones, those whose role it had been to carry tradition. It felt like a group of people left without the promised spiritual and material inheritance, recalling their sweet, tender past.

The information I was able to obtain in this meeting concerned two issues. One was related to the fact that many people were taking this transition as an opportunity to leave the organization because they were offered a high severance payment. Those who were staying were mainly the ones who were not yet able to get the desired severance pay. The staff's main struggles had been with their former parent organization, The Jewish Agency for Palestine, regarding the conditions of retirement. It was not yet clear who would stay and who else would leave. The nationwide network even staged a demonstration near the Parliament, but without results. The whole country was in depression after Rabin's murder, and no one paid attention to them.

The other aspect I heard about touched the domain of their clients. Staff were experiencing crises of various kinds, both of schools and of individuals, mainly involving threats and attempts of suicide. While talking with them I could resonate within myself to many nuances of loss, mourning, and death. However, I got the impression of a highly professional, responsible group of people in spite of the gloomy, passive atmosphere of their work situation and of our meeting.

One critical event that I was told about touched on aggression, anger, and acting out. It was an enactment within the subgroup of social workers. The leaving chief social worker appointed another woman to replace her in management. This gave rise, within the subgroup of social workers, to an instant wave of strong anger, which was directed towards the new woman. The rebellion succeeded and she stepped down. Nobody else among the social workers dared to take up the role. They tried to cope with the problem, seeing it as an internal issue within their subgroup, rather than an event that was enacted on behalf of the whole organization. As a result of their working through the issue, a new social worker volunteered to become their head and to join the management, leaving the rejected woman in much pain, and the whole staff in a traumatic state. Such a thing had never happened to them before. Through all this episode, management could not say or do anything about this crisis situation. They felt paralyzed. This event had a great impact on the whole unit.

The management group and myself decided on a one-day organizational review that would involve all members of the "station." The purpose of this review, in the format of a large group, was to enable the organization as a whole to draw a common picture of its state and to gain a better understanding of how they got there and where they were heading to. It was clear that unless all staff, including management, were able to make sense of the transition; to analyze the current reality, both overt and covert, internal and external; to decide together what needed to be changed; and to generate ideas and emotions together, they would not be able to support and facilitate the transition which they had to undergo in order to survive. It was also clear that the issue they were facing was a systemic one. This is why the whole system was called to review its situation (see, e.g., Bunker and Alban 1997). The other reason for intervening on the level of the whole unit was my estimation that management had tremendous difficulties in taking up their leadership role. I thought that in order to wake up from their depression and passivity they needed to experience the difficulties of their people. I hoped they would be able to add their deep understanding to what they heard daily, formally and informally, from individuals in the organization. I hoped they would be able to understand where their authority stemmed from and succeed in regaining it. I also hoped that the other staff members would be able to regain their own individual authority. As Miller (1993) noted, "Long-term solutions to the problem of managing change cannot . . . depend on manipulative techniques. On the contrary, they must depend on helping the individual to develop greater maturity in understanding and managing the boundary between his own inner world and the realities of his external environment."

They could not join me in designing the day of the review. They wanted me to do it all by myself. This was just another aspect of their depressive mood. We agreed to meet on the morning of the event, one hour before the review, in their offices, the space that the event was to take place.

THE ORGANIZATIONAL REVIEW: SECOND MEETING
WITH THE MANAGEMENT

I was the first to arrive. Shortly later came the chief social worker, the new member of management. The director and the chief clinical psychologist came only half an hour later. We were not left with much time to prepare for the event. I felt left alone, and I wondered whether I was sharing this experience with the staff. I was asked to wait in the director's room, the largest room in the clinic, which as I later learned, was the place of staff meetings. Before the transition they used to have a large, special hall for their staff meetings. I began to realize that through the transition they had lost not only staff members but also parts of their physical territory. On the walls of this large empty room hung framed pictures of the dead founders and former leaders of the organization. History was there, right in front of my eyes, told by the silent walls.

When we at last got to the point of talking about the event, the picture became even more complicated. I had realized that the group of people I was about to meet was composed of members who during the transition were identified as being paid for their work by three different formal authorities. This had to be a temporary arrangement, until things had settled down, but it also was the source of much confusion, anger, and despair. I was warned by the director that I would hear many complaints about money issues. He told me that he was too tired to deal with them. I tried to lead our short discussion towards the primary task of the review. They agreed on a definition of the task: to explore and analyze together the current reality; to see what needs to be changed; and to review their mission in the light of the transfer to the Ministry of Education and Culture.

I realized that during this short planning meeting I was bothering with the breaks, and with the food that should be served during these breaks, almost more than I was dealing with the design of the formal meeting. I could not understand why I was more preoccupied with the "informal" time of the event than with the "formal boundaries of working toward the task." One could say that I was dealing with the transitional space (Winnicott 1971).

One could also say that I had an intuition, a blind one, without any corresponding concept, as postulated by Bion (1977). Since my experience of the outside was so confused and overwhelming, I had to listen to, and rely on, my inner signals, even though consciously I could make no sense of them.

I came out with a timetable that was a set of empty time slots; I did not know how and with what they would be filled. I knew only about the breaks. On several occasions I asked who would take care of the food during the coffee and lunch breaks. In the end they told the administrative assistant to organize the meals with some help of the participants. I was relaxed to see the problem being taken care of, but I still did not know why I had become so anxious about it. When I was asked about it by the management I was only able to say that, in organizations, informal shared time is as important as the formal time, and that attending to the well-being of staff is always a loaded issue, especially in times of transition when the ground is felt to be so unstable. It was clear to me that through my dealing with boundaries and breaks I was containing something important for them, only I did not, at the time, know what it was.

THE REVIEW

A group of 30 people, men and women, were sitting in the big, long, and narrow office of the director, around a big, long, and empty table, with their backs to the walls. The table was brought in just before we started. I looked at the table being brought in, and thought that it could have at least two purposes: the first was that the empty space in the middle of the room would be filled, so as not to leave room for a void, or for a transitional space to play in; and the second purpose could be that because of that big table no one could leave the room, unless he was sitting right near the door which was located in the far narrow end of the room. I could, of course, raise a third hypothesis which went along very well with my special attention to their nutrition, namely, that such a big table could carry a lot of food for the breaks. At that moment I could not know how this last hypothesis would go along with the two others. I was very alert to the fact that if anyone would have wished to leave the room while we were still talking, many others would have to stand up and leave the room in order to create a free path for him to leave.

The first part of the review was dedicated to individuals voicing their feelings about their position in the organization and their expectations from

this organizational review. In the forthcoming discussion I will try to cap-
ture, as much as possible, the exact wording, as well as the sequence of the
meeting, in order to enable the reader to listen to both conscious and unconscious tones and undertones of the process and its content.

The event started with a short opening speech by the director who related the management's considerations which led to inviting an outside consultant who would lead the organizational review. He asked me to introduce
myself and handed the group over to me. After introducing myself, I presented the program for the day and the goals to be achieved. I also mentioned that the goals were defined by management at a meeting with me.

The first to talk was the social worker who had been "forced" to step
down from the management after the "mutiny." She said that after a long
period of great pain it was about time to start being connected to the future. She expressed the feeling that she was belonging to an organization
she did not know and could not recognize anymore. She added that she
felt like being betrayed by her own family, a family which she felt she knew,
but then realized that she did not know. Someone angrily raised the question whether it was compulsory to participate in the review. It was as if he
pointed right from the beginning to the danger of openly expressing anger,
rather than shutting one's mouth, keeping one's anger suppressed. He was
also raising authority issues right from the start: management's authority,
my authority, and the authority of the individuals. I could sense a lot of
tension in the air, and started to think that in this organization taking up
a role and using any kind of authority—leading or managing—was dangerous. Authority was confused with being aggressive. I thought of the
depressed management, who found it so difficult to lead.

There was a silence, and since no one from management related to
this response I decided to intervene by saying that I did hope that they
would be able to create and develop an atmosphere that was safe enough
to talk openly. Someone else said to me aggressively that this "command"
of mine to be spontaneous was a difficult one to follow. He then continued, bitterly, that the situation had to do with control and change. How
was it that they, a group of professionals, experts in change of others, found
it so difficult to transform themselves? Feelings of helplessness did not allow
them to find within themselves a possibility to master any part of their own
changing reality, he continued. His voice became softer and quieter, almost difficult to hear, as he came to the end of his intervention.

After that, M, a woman in charge of the consultation services to the
boarding schools, said that her personal aim at this review was to be able

to influence. In her eyes the organization was a "transit station" with a long past but with "little future"; an organization with an almost unseeable future. In the past, all staff were equal, but now they are so different from each other. They worked under different contracts, some of them were waiting for the most convenient time to retire, and others were holding strongly to the old habits. She would like the group to be able to take responsibility for their reorganization. Only if they could do this would they start to master their future.

N, a woman social worker, the oldest and most senior staff member, albeit without fulfilling any managerial role, was waiting for her legally "right time" to retire. She talked about her seniority and about her anxiety as to the future. I thought about her and her role in the system as the person who could see the farthest back into the past, because she carried and contained all the memories, and could carry, among others, the tradition to the future. She was the first to create the link, in this meeting, between Rabin's assassination, the turbulence the country was going through, and the earthquake their system had been experiencing. She concluded her comments by saying that everything was different now and they would have to prepare themselves for a different future. I must say that she sounded like a prophet of wrath.

S, a psychologist, described her state of emotional confusion. On the one hand, when meeting patients and clients, it felt as if nothing had happened, she enjoyed her work like in the old days. On the other hand, she had not received her salary, the first payment that was to come from the Ministry of Education. "How do we look?" she asked. "How much strength do we need? Are we resilient enough?" She was the first to mention the depression of the management by saying that the director was tired and the chief psychologist did not maintain her strength anymore. "If people and money would not be here, what could we do, how would we regain resilience, energy, hope, and faith?"

Another woman, G, said that fieldwork was very hard for her. Coming to work had become more and more difficult. No one cared anymore about the parking lots for the staff. She realized that each day she was parking her car in a more distant spot. The future seemed to get more and more distant. Would they be able to bring the future closer? L, a senior male psychologist, said that he had been waiting a long time to be able to talk about the crises. For him, being a staff member involved undergoing many crises on different levels. He related to crises from the past, crises that had been woven into the life and history of the organization, personal, professional, and ideological ones. "The whole picture is very unclear. It involves questions about the essence of work, it involves feelings around losing many close, old friends, and it involves strong fatigue. My only request is that they will leave us to do the work we know and let us think quietly." Who were "they," I thought, but did not say anything.

M, another male psychologist, who was very quiet and depressed until that moment, said that for him the station was always a place "of slippers," meaning a place where one could feel comfortable, at home. "But now, every time I come to work I see digging and new construction projects all around our place. Who has the strength to dig? I am afraid, though, that without digging we will not be able to move on. We have to dig into ourselves."

N, the social worker who had recently joined management, was the first member of management to participate in the discussion, even though she said that she had wished to speak from her former position as a regular staff member rather than in her managerial role. Again I wondered how difficult it was to take up a leadership role in this organization. Did it mean that they would not be able to wear, all of them together, the same comfortable slippers and maintain this wonderful feeling of a tightly knit family of siblings who lost their parents? She mentioned the fact that they were now the property of the government. They were owned by the Ministry of Education and Culture. Their future was connected to the future of the nation. She asked if there was a way to make reality easier. The station used to be a very attractive working place; would they be able to bring back this attractiveness? Would they be able to regain their old way of communicating with each other, she wondered.

E, a female psychologist, could not look into the future. She was still very much occupied with the past. She said they all shared the illusion of the station being a family, a group of people who built a home for themselves. They all closed their eyes to reality, she said. The omnipotent belief that "nothing wrong could happen to this family" was a defense that did not help them to face transformation. The shared illusion exploded, and now pain and disappointment forced them to fight for themselves, for the respect for their working place. "The fight did not last long. Depression won, people felt helpless. They left. After all, it is only a working place," she said.

I thought to myself that their everyday work was about trying to build new and warm inner homes for all the children and youngsters they treated. And now all of them were experiencing the same feelings like the rejected, neglected children, in need of a safe, warm home.

The next to speak was A, a female intern. She was the first intern to take the floor. I found out that until this review the interns had never before been invited to take part in the discussions about the transition. They were kept out of the inner circle of "adults" and protected from horrors of life. They were not the only group in need of protection. The "adults," their supervisors, had to protect themselves from being perceived as helpless and depressed as they were. This woman expressed her feelings of not really knowing how much she and the other interns were part of the system. They did carry a lot of responsibility in the work they were doing, and in that respect they were regarded as adult professionals, but when it came to sharing painful information they were still regarded as ignorant youngsters. "By what standards were people appreciated in this organization?" she asked. By asking that, she was able to voice a difficult feeling that was part of the transition and perhaps also part of the previous, remembered crises, a feeling that was shared by many people in the room. These feelings had something to do with how one knew one's place and value in the organization.

The last sentence before the coffee break was uttered by S, who had spoken before. She said that they needed to reach clarity as to whether the network was to be closed down or handed over to the Ministry of Education and Culture. She had the feeling that they were operating under the hypothesis that they were closing down. After the intern "child" spoke, from the boundary of belonging, a remark could be made about being part of a nationwide organization.

It was very important for me to adhere to the timetable; thus we broke and I did not say much at the conclusion of this segment of the meeting. I must admit that I did not know what to say, even though much went through my mind when listening to the participants.

FIRST BREAK

The big table was immediately loaded with cakes, sandwiches, sweets and cookies, soft drinks and hot drinks. During the break people were very active, talking to each other in a lively way that projected a different atmosphere than the one that prevailed during the formal work session. I was left alone; no one approached me. I kept to the timetable and invited them back to work after 15 minutes. The food remained on the table while we continued the review.

THE SECOND PART OF THE REVIEW

I opened the second part of the review by sharing with them my impressions of the morning. I described the mood, I combined the metaphors that were used by them or crossed my mind into a narrative. I reminded them of the important mission of their nationwide network, namely, to save and rescue children. I said that I could understand that they as an organization could identify very much with the rejected, lost children they had been taking care of for so many years, and that they had found themselves in the same situation, being lost and deserted by their own parents. That the transition from the Jewish Agency to the Ministry of Education and Culture was experienced by the organization as a narcissistic injury inflicted to the core of their very being, to the core of their personal, professional, and institutional identity. I also said that on top of the difficult conditions they got from their new "owner," there was also the fact that the Ministry of Education and Culture was a much ridiculed governmental body and by being so, it was an unstable political entity which elicited a lot of anxiety around the mission of saving and rescuing rejected and lost children. A thundering silence prevailed in the room for a while.

Then O, a female psychologist said: "What will happen to the children if the parents are sent for one year on probation to a new foster home?" The agreement between the Jewish Agency and Ministry of Education and Culture stipulated that the network would continue to work the way they were used to for one more year after the handover, and then the Ministry would decide how they wanted to operate the system. They felt that they were on probation, not knowing what would happen to them, whether they wanted to continue the way they were used to, or not.

Z, the chief psychologist, the other female member in management, said: "We have to continue to play the music . . . this tune cannot be stopped." These were two lines of a patriotic song known to everybody in the room. She then shared with the group her anxieties about youngsters committing suicide in the boarding schools. She spent the whole summer worrying about the high numbers of suicide attempts and actual suicides. She knew that it was related to the crisis they were undergoing because of the transition. She continued to say that she had been feeling like someone who is losing her freedom, losing the right to speak up. They could not choose whether or not they wanted to belong to the Ministry of Education and Culture. No one had asked them, as people having a leading role in the organization. She herself felt this to be a very shaking and insulting experience. She also reminded the group that during this last year of difficulties they lost H, their colleague who died of cancer. For her, depression and death were with them all the time. "But," she added, "there are also very stable elements in our life; our mission is very clear, our ideology and professionalism are clear and stable. So what has changed?" And she answered this herself: "Work and economic conditions have changed, our ways of organizing ourselves internally have to change. We were forced to become an organization with subgroups. We cannot feel equal and united as before."

D, a young female intern, intervened immediately after Z stopped, as if not to leave room for the group to feel and reflect upon Z's words. Perhaps it was too frightening and not clear enough how they could hold on to hope and strength, when the leaders were preoccupied with death and loss. D said that it was an illusion to regard their place as being the same station like before the turbulence. People left and their names were still mentioned every day. She missed them very much. Subgroups were divided because of pay conditions. They needed to find a way to build a new group. I thought that she was voicing the wish to be reborn, to start things afresh, not to remember the pain.

After that, all of a sudden, I could hear very emotional talk from many corners of the room. It was talk about the cafeteria. The atmosphere in the room resembled that of the break. It was difficult for me to understand what was going on because people were talking together at the same time. I let it go for quite some time before I was told the story: All these long years, since the network was established, they used to have a cafeteria on the same floor of the building as part of their office space. During the transition they lost part of that space, part of the territories that belonged to the old organization, the Jewish Agency. The cafeteria, the common area for informal meetings, was the most important space in the station, they said. This was where they used to meet each other and do the most important talking and negotiating. This was where they had been used to "close" work issues among themselves. The cafeteria used to serve drinks and food of all kinds, from snacks to more substantial food. By losing the cafeteria they lost the most important meeting area of the organization. It felt like losing a transitional space, a place to share, to plan, to imagine, to quarrel, to support, to create, to rest, to nourish each other and to be nourished. They were operating as a professional team mainly in an informal place, on informal time.

Hearing all of this reminded me of "The Northfield Experiments" and the "Discovery of the Therapeutic Community" (Bridger 1990). I could not help thinking of this organization as a community losing its transitional space, losing its links with the former facilitating environment (Winnicott 1965). This could very easily be felt as a traumatic experience in the life of individuals as well as in the life of organizations. Things became clearer to me, as I had started to make sense of my hunch, from the beginning of the day, regarding their food and break arrangements. I had also absorbed some more important information: In the old days, when they had the cafeteria, the door of the administrative assistant's office was always closed. The administrative assistant was a very "difficult" woman, they said, who kept herself apart from them. She was one of those who left with high severance pay. The new administrative assistant, a woman who had worked there before, and was promoted to this job, bought a big samovar, put it in the office, and began to leave her door open. One of the retiring persons bought mugs for all the remaining staff as his goodbye gift.

Only after the whole staff became aware of this loss, in addition to all the other losses they had recognized already, could they begin to take responsibility for their present situation and relate in a different way to the future. It felt like a dramatic shift in the review process. The gloomy, passive atmosphere changed to a more vivid mood. To me it was very salient that the director had not yet said a word since his opening speech at the beginning of the day. He appeared to be holding on to the depression very strongly. I realized that I too was more active in the second part of the day. I could listen in a different way to what they had to say. The understanding, created by the context of the "cafeteria" event, shed new light on their deprivation.

Knowing that we only had another half day encouraged me to add to my role the aspect of an interpreter who was relating the personal, individual contributions to the organizational reality, not only by describing my understanding of the emotional state of the group, but also translating it into a "what to do about it" mode. How could they implement the conclusions from their understanding of the transitional process? I remember thinking that they had left an autistic phase of the transition and had moved on to the phase of beginning to understand the environment they were living in. My interventions began to link even more the meaning of their words to the organization as a whole including its outside environment.

The issue of the cafeteria elicited metaphors like "weaning" and "Whose responsibility was it to feed us?" Was it in their hands or did they wish to remain dependent on the former system? Questions like "How are we going to nourish ourselves?" were formulated very clearly. The status of the organization within the new system was beginning to be examined. They had realized that until that moment they had never tried to meet anyone from their new parent organization. Only one person in the headquarters, the director of the entire network, handled all the negotiations with the Ministry of Education and Culture, and no one ever dared to question him. They had not even planned to demonstrate their professional and organizational capabilities to the Ministry. Did they have any influence at all on the way the newly born organizational arrangement would develop? Did they have any mastery of the situation? Could they feel free to express their wishes, needs, plans, beliefs? Could they negotiate about anything? What was the meaning of autonomy? What was the meaning of authority?

Then O, a social worker, said that she felt the process was going too fast, and she still had many technical questions to ask. She said: "I need more time; it will take a while for the egg to walk on its legs" (i.e., for the egg to become a chicken). M, the head of consultation services said: "Now we need the cock to come!" This response invited the group to explore, with my help, the possibility of anybody taking the role of "a cock" in this organization. Perhaps they could conceive of themselves and operate effectively only as a bunch of young chickens, without a cock fighting the other cocks, fighting those who wished to take his place and role. Again they touched on issues of power, authority, responsibility, and aggression. The male psychologist, M, who had talked before about being a "station of slippers," said: "I feel behind all of you."

There were only a few minutes left before the lunch break, and all of a sudden, out of the blue, the social worker, who was forced by her colleagues to leave management, burst out in a very aggressive way against S, one of the psychologists who had expressed her wish for everything to remain as in the old days. That was the last thing that happened before the 45-minute lunch break. Everybody was stunned. I was trying to make sense of this attack as an expression both of anger and of the wish to emphasize "who was doing the dirty work." The social workers saw the psychologists as those who are sitting in the ivory tower of pure psychotherapy, closed in their offices, in contrast with the social workers who get in touch with the community by working in the field.

LUNCH BREAK

Lunch break was very lively. A lot of food was again put on the large table. Small groups were gathering, talking very enthusiastically. This meal took place in the light of awareness of the missing cafeteria. Many people approached me to talk about various things, unlike during the morning break, so I did not feel deserted as I did then. There remained still two "slots of time" ahead of us: looking at the future and the conclusion of the review.

THE AFTER-LUNCH SESSION

After lunch, the social worker who had become angry apologized for the violent expression of her feelings. She addressed S personally and said that she could not understand what her outburst meant. Of course she was very hurt by what had happened to her, but she could not tell why she had attacked S. She only knew that she was carrying an emotional overload.

I asked if they could make sense as to where, in the organization, this emotional overload belonged. L said that in this place it was frightening to be active but also very frightening to remain passive. Others said that it was difficult to express anger. The anger was about something that paralyzed them. I asked if they felt that they could have any influence on and mastery of anything. Were they united? Were they partners? Or, did each of them carry the load alone? They could tell that in the course of the transition their feelings of cohesion had been weakened. I thought they were on the verge of talking about their leaders, but they did not yet dare touch the issue of leadership in the organization. Was it so frightening to touch a weak management?

As if she was listening to my thoughts, Z, the chief psychologist, said that as far as she could understand, they were talking about power. This was a problem that also bothered her personally. To what extent was it permitted in this organization to be active, to push ahead, to take new initiatives, to build bridges to the future? Were they dealing with conservation or with development? She could hear her colleagues hesitating about the right timing for new developments, but they all had to face reality. Why not develop a strategy and move slowly forward taking into account the difficulties and their ability to overcome them?

Suddenly management was not one silent front anymore. I could sense cracks in the solid wall. Z continued: "In such an unclear situation, one needs to move very cautiously." She said that recently she had been feeling very fragile; she could sense much unspoken criticism against her and against management as a whole. "Why did everything get so magnified in the eyes of people?" she asked. "Why is it that every little step forward was experienced like a giant leap?" I related these comments to the fears of losing the old sense of the group's self. If only they could find a way for everyone to agree, they could risk moving on.

E said that the most frightening thing was to lose their tradition, to become something else. How could they develop a myth that there was unity between all of them? I raised the question of the differences between the two professions in the station. They were all talking as if there was no differentiation between the professions and between the roles in the station.

They all started to relate to an event that happened amongst the social workers. They could observe that the anger against leadership was expressed by this group who refused to obey the last instruction of their former chief when she wanted to determine who was to follow her. They could vent their aggressive feelings only when seeing the back of the departing leader. They had to assert their power opposite their "dead" leader and her chosen replacement, even at the price of hurting one of their own. By doing so they broke one of the most important unwritten rules of the organization.

As we were heading for the concluding part of our session, after all these difficult realizations, the director took the floor. There was deep silence in the room. The only noise one could hear was the "noise" of listening. He said: "I was listening very carefully all day. I can identify with everything each of you said. All this year I had the image that we were thrown into a deep stormy river that no one could master. The only thing one can do in such a flow is to hang in there, to let oneself be driven by the water and see where one gets to. The conflict within the group of the social workers showed me that a lot was going on underneath the surface, but I did not know what to do with it. Now I do not feel so passive anymore in this stormy river. I think we got to the shore, wounded, sick, and hurting on many counts: personal, economical, organizational. We need to rise and start walking. We need to relate to the change, but also to take care of ourselves, having not much energy left. We have to survive. We lost old friends, new people have joined us. The newcomers have energy and ideas. We have new members in the management. Now let us start the new year allowing ourselves to accept the new reality. It will not be easy but we need to start exploring our new territory. Let us bring our old tradition into the new territory." There was absolute silence in the room.

It felt like the right thing for him to say, and it felt good for the group to hear these things at last. The only clue I had that something was wrong was the fact that he, as the director with whom I planned this day, knew the timetable and nonetheless allowed himself to cross this boundary of time without relating to the fact that he was doing so. I decided not to touch the issue, and got rid of my disturbing thoughts. I knew we were on the verge of concluding the day, and after all he had at last addressed his people. He had invited them to go along with him. These were the words, and maybe I was not the only one to have doubts. I was aware of the fact that a very important part of the task had not been addressed. No exploration of the new reality and the new father organization had commenced.

THE LAST PART OF THE REVIEW: CONCLUDING THE DAY

V, an intern who had not yet spoken, was the one to open this part of the discussion. She said that she was oscillating between feelings of being part of the organization and of not belonging. The transition from the cafeteria to the samovar was for her the signal of the most meaningful change that had happened to the organization. The steps toward the future should be small ones. The decisions as to how best preserve the feeling of belonging together was for her the main step toward the future. Someone else followed her, pursuing the same theme about different losses for different people. I felt tense. I was afraid that like a posttraumatic group they might go on and on with the same complaints. I was worried about the task of this last part of the day and the overall primary task of the review. I decided to intervene. I mentioned that we were approaching the last part of the day and that they might want to look at the process they had undergone and try to consider how they would carry themselves from now on in a more caring and constructive way.

People began to suggest ways of working toward their future. They considered, for example, such issues as preserving the past versus changes they need to effect, as well as the totality of their decision-making process. Their experience had taught them that whenever decisions were not made democratically, the staff would not go along with them. They insisted again that their ability to feel together as a group and close to each other was the most important aspect of their common life.

I became aware that I was beginning to feel angry. I thought that they were indulging in the idea that they had a single, shared fantasy. I was able at that moment to formulate a hypothesis about the myth of being together as a social defense that neutralized the possibility of differences and differentiation, and paralyzed any possibility for leadership to emerge. I was afraid that if I did not share my hypothesis with them, we might end the day having filled all the time we had at our disposal with the same material. I was also afraid that my anger would fade out, like the leadership in the organization. So I told them what I was thinking and added that I could feel that the pressure from the outside was very strong and the wall surrounding and protecting them getting thinner and thinner, with the risk that it may eventually crack. I could see them focusing all their attention on keeping themselves together, holding hands, not letting anyone do anything else with his or her hands. This attitude would demand a very high price.

They became very silent. Someone suggested they should work on the way they were listening to themselves as a group and to each other as individuals. I told them that it was very difficult for them to leave the autistic position and start listening to the outside reality. Their leaders were not able to sit on the boundaries of the organization, maintaining links with both the outside and the inside reality. The system had swallowed up its leadership to prevent the destruction of its autistic defenses. They had surrounded themselves with an impenetrable wall.

In response to my intervention, they wanted to find out whether they could allow the "space" that had been created with me to continue. I wondered if this reaction was the result of them feeling dependent on me, the person who came from the outside to help them out, or if it was a sign of their ability to contain the function of the cafeteria, but this time as an internal transitional space rather then a concrete one.

The clouds were dissolving. I could hear voices of people starting to mention constructive steps that they had undertaken during the past year. Others talked about plans for the future. They started to explore together who was to give the blessing for any kind of initiative. Some said that they would like the management to be less passive and more leading. One of the interns said that she had imagined a very long meeting with the management where they would receive all kinds of decisions for moving forward, thus creating contact with the Ministry of Education. Others called on the management to get out of their depression. The most senior woman on the staff, N, said that she could recall a meeting they had with the director of the department in the Ministry of Education. At that meeting, the department director had said that this year they would continue to work in their own way, but be observed by them. After that, the director would talk to them. She added: "We have to get ready to meet him. Let us prepare ourselves and invite him." Her words were accepted by the staff. It

was agreed that without them getting to know the system they were joining, they would not be able to survive and to carry on their important
mission.

TWO WEEKS LATER: LAST MEETING WITH MANAGEMENT

The last meeting with management took place in my office. We shared
impressions. They talked about the staff's responses to the review. They
said that people could relate to the heavy load they were carrying, a load
that had become an organizational load rather than a personal, individual
problem as before. They told me about a long management meeting they
had held two days after the review, at which they realized how much all of
them identified with the role of orphans, parenting each other, without
grown-ups around. They found it very difficult to assume the role of grown-
ups, who would know how and where to lead their staff.

Our work in this meeting concentrated mainly around their difficulties in taking up their role as management. We examined their personal
situation in the organization in relation to the role they were fulfilling. One
main understanding they had after the review related to the way roles were
given to members of the organization. They had realized that role distribution was more affected by personal considerations than by organizational,
functional ones. They said that they were always led by the wish "to keep
things quiet," instead of confronting their staff with managerial decisions
that might give rise to anger and negativism. They understood that the case
of the social worker's representative in management was an example of this
unconscious attitude; that M was the victim they had chosen, unconsciously, to protect them as management against expressions of anger by
their staff. They also understood that in order to fulfill their roles and tasks
they needed to develop their capacity for teamwork.

In this meeting they were able to talk about the differences among
themselves. The director said that it was very difficult for him to function
differently in his roles as a psychologist and as Director. Z, the chief clinical psychologist, said that she was perceived as the most "bossy" person in
management. She was the one who told people what needed to be done.
She was trying to look into the future and find a way to continue. This was
why she absorbed so much anger during the last year and was the target of
criticism and complaints from the staff. Z asked her colleagues in management to take on more authority and not let her be the only "bad" one. As

a result of the review and our meeting following it, management agreed to continue working, on an ongoing basis, on their roles and tasks, with the help of a consultant.

The night following our last meeting, Z was hospitalized for a heart condition. She could allow herself at last, as she said later, to get things off her chest, and to take some rest.

CONCLUDING REMARKS

When thinking of my work as a consultant, meeting organizations in various ways, at different angles, meeting them through individuals, through small groups of people or large ones, it is always clear to me that my insights are being built anew each time and are informed by three main theoretical sources: psychoanalysis, open systems theory, and group relations theory. When approaching a new project, I can often see how understanding emerges with the help of concepts derived from these theoretical perspectives. To give an example: I can see a project through the frame of a "Group Relations Conference," using elements of that frame in order to gain a better understanding of the situation, such as the project I described in this paper, a review that took place in the format of a "large group." I can also think of major transitions through the model of a "Group Relations Conference," and relate to the aspects of moving from one setting to the other, facing and forming different kinds of relationships and relatedness to various groups within or outside the organization. But I would not have been able to speak about "within" and "outside" without relating to the notion of boundaries, which is a central concept of the open systems theory (Miller and Rice 1967).

> The existence and survival of any human system depends upon continuous interchange with its environment, whether of materials, people, information, ideas, values or fantasies. The boundary across which these "commodities" flow in and out both separates any given system from, and links it to, its environment. It marks a discontinuity between the task of that particular system and the tasks of the related systems with which it transacts. Because these relations are never stable and static, and because the behavior and identity of the system are to continual negotiation and redefinition, the system boundary is best conceived not as a line but as a region (Lewin, 1935; 1936). That region is the location of those roles and activities that are concerned with mediating

relations between inside and outside. In organizations and groups this is the function of leadership; in individuals it is the ego function. [Miller 1990, p. 172]

Needless to say that all these analyses are grounded in the theory and understanding of psychoanalysis.

The advantage of basing the understanding of organizations on these three theoretical frameworks—psychoanalysis, open systems theory, and group relations theory—lies in their focus on the interrelatedness of the individual, the group, the organization, the community, the society, and the environment as such. In that sense they embrace at once the widest possible range of invisible and unconscious influencing factors (e.g., the missing cafeteria, or even Rabin's assassination), and the innermost feelings and emotions that lie at the core of humanity: envy, anxiety, attachments, hatred, helplessness, power, authority, hope, and so forth. Of course this list could be much longer, and each of us carries their own list.

In this case study I have tried to show how the "big things" and the "small things" go together, hand in hand, in the process of a major transition, but also in the life of individuals and organizations, in general. After all, what is development if not a continuous transitional process? Individuals and groups interact in order to give meaning to their experience, thereby developing mechanisms that can protect them against the unknown. These mechanisms, often unconscious, are deeply rooted in the life of the organization and are threatened by the prospect of change (Miller 1990).

Let me relate more specifically to some theoretical problems that I have been left with after this one day of organizational review. The people I met with were a "work group" with a common task (Bion 1961). I considered it a "large group" because of the number of people engaged in the same task in the same place at the same time. Given this, the underlying dynamics were that of a large group in the sense that in such groups people feel less intimacy; they are "faceless," as it were; they lose their individual identity and are driven more into the undifferentiated mass of the organization. Members of such groups usually address the group as a whole rather than as individuals. Contacts and bonds between individuals are swallowed up by the big "black hole" of the large group. Envy and aggression are more easily expressed because of the illusion of not looking into the eyes of the individual. Projections are easier to make and less restrained. I keep asking myself: What happens to the task when such powerful "large group" dynamics are operating? (Turquet 1975). Is it at all possible to review the task of an organization with the participation of all of its members?

Perhaps it is possible to review the task in a "large group" if the participants can contain the notion of the "small group" within themselves, and at the same time maintain the feeling of being part of the "large group," of the whole. The small group is a metaphor for home, for family. Even though everybody knows that emotions inside the family are not always consoling and forgiving, the members of the small group do not lose the feeling of intimacy and containment. The awareness of the existence of walls, rooms, and links is preserved in a small group. Because of the small size of the group, one's "face" and identity do not get lost. Each member is conscious, to a lesser or larger extent, of his contribution and his value. When becoming part of a "large group," the members of the small group tend to lose their feeling of having a home. They lose their self-confidence. They become less creative, less productive, and less communicative. They tend to be overcome by feelings of anxiety and loneliness.

The function of the "large group" in pursuit of its task could be more effective if its members would retain in their personal and group memory the experience of belonging to a "home group." In this way, the notion of "home" would be permeating the "black hole" of the large group, bestowing on it humaneness and enabling it to undergo the transformation to a less persecuted and persecuting entity. Such an entity would be more communicative and open to the world; it would facilitate the flow both inside out and outside in; it could assimilate transitions as part of its development and not only focus on its traumatic aspects. This raises an important question for the practice of consultation: Is there a way for the consultant to help the large group—the organization—to maintain the notion of a "small group" while functioning as a whole?

Another problem I encountered is related to the fact that a "large-group" event in a Group Relations Conference is a "here-and-now" event by definition and the "here and now" is itself loaded with many experiences of the moment. It is more difficult to look into the past, or to try and see a common future when the present reality, overt and covert, is so vividly present. A review is a task that invites exploring the links between the "there and then" to the "here and now." The role of the consultant in that respect is to facilitate these linkages in order to gain a better understanding of the transition.

I would like to conclude with the words of Eric Miller from the preface to his book *From Dependency to Autonomy* (1993):

Change, even when intellectually people see it as necessary and desirable, always arouses anxiety. It is part of my task as a consultant to contain some of that anxiety so that members of the client system are not crippled by it. Dependency is therefore not to be avoided but managed during the phase in which they are gaining greater mastery over the situation. My task is then to become redundant. The intervention will be successful if clients have transformed the dependence on me into fuller exercise of their own authority and competence. [p. xviii]

These steps leading from dependency to autonomy have to be gone through not only by the client system but also by the consultant. In each consultancy project, the consultant has to walk this road again and again.

REFERENCES

Bion, W. R. (1961). *Experiences in Groups and Other Papers*. London: Tavistock Publications.
———— (1977). *Caesura: In Two Papers the Caesura*. London: Karnac Books, 1989.
Bridger, H. (1990). The discovery of the therapeutic community. In *The Social Engagement of Social Science: A Tavistock Anthology*, ed. E. Trist and H. Murray, vol. 1, pp. 68–87. Philadelphia: The University of Pennsylvania Press.
Bunker, B. B., and Alban, B. T. (1997). *Large Group Interventions*. San Francisco: Jossey-Bass.
Lawrence, G. (1997). *Centering of the sphinx for the psycho-analytic study of organizations*. Paper presented at the ISPSO 1997 Symposium, Philadelphia.
Miller, E. J. (1990). Experiential learning in groups I. In *The Social Engagement of Social Science: A Tavistock Anthology*, ed. E. Trist and H. Murray, vol. 1, pp. 186–198. Philadelphia: The University of Pennsylvania Press.
———— (1993). *From Dependency to Autonomy*. London: Free Association Books.
Miller, E. J., and Rice, A. K. (1967). *Systems of Organizations*. London: Tavistock Publications.
Turquet, P. (1975). Threats to identity in the large group. In *The Large Group*, ed. L. Krueger. Itasca, IL: F. E. Peacock Publishers, Inc.
Winnicott, D. W. (1965). *The Maturational Processes and the Facilitating Environment*. London: Hogarth Press.
———— (1971). *Playing and Reality*. London: Tavistock Publications.

Complexity at the "Edge" of the Basic-Assumption Group*

RALPH STACEY

Since the theme running through this volume is that of learning from experience, I feel it appropriate to start this chapter with a few brief remarks on some of the key experiences that have led to the views I present here.

For many years I practised first as a corporate planning manager in a large construction company and then as a strategy consultant. In early 1989, I completed the draft of a book in which I tried to set out the sense I was making of this experience (Stacey 1990). The key question in my mind at the time had to do with why the long-term planning activities I and others had been involved with seemed to bear little relationship to what subsequently happened; and why, in the light of this experience, we carried on doing it. It was at about this time that I was asked to take charge of a unit on the MBA programme at the business school where I now teach—the person who had designed this unit had left before it began. The major part of the unit was to be a weekend group relations training conference directed by Eric Miller. I am embarrassed to confess that, while I had heard of the Tavistock Institute, I had no idea what a group relations training conference was, nor had I read any of the literature that had been flowing from that institute for decades. I saw my role at this event as mainly one of taking care of the accommodation arrangements and generally seeing what

*This paper owes much to intense discussion with colleagues Douglas Griffin and Patricia Shaw. I am also grateful for the comments of Eric Miller and the editors of this volume on an earlier draft.

it was all about. My experience of that first conference ranged from shocked disbelief to somewhat hysterical amusement and I left it profoundly affected without knowing why, except for a strong feeling that what I had experienced had something to do with the nagging questions about the efficacy of long-term planning activities in organisations. I became an addict and took part in many more group relations training conferences, and what I learned there has come to have a major impact on how I am now trying to make sense of life in organisations.

A few months after this first conference, again by chance, I came across a book called *Chaos: Making a New Science* (Gleick 1988). As I struggled to understand this, I experienced a growing sense of excitement at what I dimly perceived to be connections between this new science, my experiences in group relations training conferences, and those nagging questions about the efficacy of long-term planning. More purposefully, I then moved on to reading about far-from-equilibrium thermodynamics (Prigogine and Stengers 1984) and complexity theory (Gell-Mann 1994, Goodwin 1994, Kauffman 1993, 1995, Levy 1992, Waldorp 1992). Along with chaos theory these are some of the strands in the wider field of thinking known as "nonlinear dynamics".

My fascination, perhaps obsession, with groups then took the form of embarking on a group analytic training in 1992, and then in 1993 setting up a group of Ph.D. students interested in using notions from nonlinear dynamics to make sense of life in organisations. The former provided me with experiences in groups from a somewhat different perspective to the Tavistock one; and students in the latter overcame an initial reluctance on my part to thinking in social-constructionist terms.

What follows in this chapter is my attempt to tell a story about the nature of groups and organisations derived from the experiences I have outlined above. I call it a "complexity perspective," and in taking this perspective I have been attempting to emulate the integration of systems thinking with psychoanalytic perspectives which I think is one of the most important features of what has come to be known as a "Tavistock approach". However, in doing this I am interested in exploring the consequences of integrating somewhat different approaches to interaction than those found in the systems thinking employed in a Tavistock approach as well as some additional psychoanalytic and group analytic concepts. The result seems to me to be a story that is similar in some respects to the story called a Tavistock approach but differs in others: one story privileges some aspects of groups and organisations, while the other story privileges yet other aspects. In line with my more recent interest in social-constructionist thinking I do not see either of these stories as the truth: they complement each other in some senses and contrast in others, and both have something useful to say about life in organisations. However, I will not be able to conceal my personal preference for one of them.

A Tavistock approach, as expressed in group relations training conferences, is not simply a systems theory, but it is the systems theory aspect of it that I will mainly focus on. I will first outline what I understand to be the systemic concepts employed in a Tavistock approach and then point to some questions that are raised about the nature of systems by developments in nonlinear dynamics. I will then move on to a discussion of how a nonlinear dynamic perspective might also have implications for the kind of psychoanalytic constructs one employs. The chapter will end with a discussion of how different views on interaction might lead to different practices as a consultant.

THE NATURE OF SYSTEMS

The central systemic concept in a Tavistock approach is that of the open system (von Bertalanfy 1968), while the central systemic concept in the complexity perspective is that of nonlinear dynamics, already referred to. Both of these concepts, as they are applied to human action, are imports from the natural sciences—von Bertalanfy was a biologist, as are many of those developing insights into nonlinear dynamics. An open system exists by importing energy/materials from its environment across a boundary, transforming them, and then exporting them back across the boundary (Miller and Rice 1967). As it is used in a Tavistock approach, this boundary is seen as a region in which mediating, or regulating activities, occur to protect the system from disruption due to external fluctuations but also allow it to adapt to external changes (Miller 1977). The boundary region must therefore exhibit an appropriate degree of both insulation and permeability if the system is to survive: the metaphor of a membrane comes to mind. Von Bertalanfy wrote about the self-organising capacity of open systems and this was recognised by those writing in a Tavistock tradition:

> such systems may spontaneously reorganize towards states of greater heterogeneity and complexity and they achieve a "steady state" at a level where they can still do work. Enterprises grow by processes of internal

elaboration and manage to achieve a steady state while doing work, i.e., achieve a quasi stationary equilibrium in which the enterprise as a whole remains constant, with a continuous throughput, despite a considerable range of external changes. [Emery and Trist 1960, p. 85]

This is a clear recognition of the nonlinear dynamical nature of open systems. However, as I understand it, this spontaneously self-organising capacity does not occupy a central position in a Tavistock approach and neither was it taken up in the wider literature applying systems thinking to organisations. The emphasis in a Tavistock approach seems to be more on the regulatory function at the permeable boundary region than on the spontaneously reorganising capacity of the system. This emphasis implies, it seems to me, that the regulatory activity by leaders/managers at the boundary is seen as the key to changes in the system. With regard to human action it then becomes quite logical to think about change in terms of rational design and to look for what might inhibit such rational designing activity. Disorder is seen as an inhibitor which must be removed.

By contrast, consider the work of the chemist Prigogine (Prigogine and Stengers 1984, Prigogine 1996). He too is concerned with open systems, but the focus is much more on the systems' self-organising capacity than on regulation at a boundary. The focus is on the systems' internal ability to spontaneously change itself in the absence of any design. He showed in experiments with chemical systems that when the rate of addition and extraction of a particular chemical into a mixture increases to a critical point, pushing the chemical system far from equilibrium, small fluctuations are amplified and it suddenly and spontaneously self-organises into a more complex form. He called this more complex form a "dissipative structure" in that it remains in being only while the additions to and extractions from (imports/exports) the system stay at the critical level. He showed in these experiments how new order can emerge spontaneously from fluctuations through a process of self-organisation: order out of chaos. The emphasis is thus shifted from regulation at a boundary to the manner in which the system's transformation process transforms itself. Prigogine has moved away from the underlying concept of an equilibrium-seeking system, clearly indicated in the above quote from Emery and Trist, to notions of systems operating far from equilibrium where spontaneous self-organisation can produce emergent novelty. This is a major shift in perspective at a fundamental level—one that emphasises the creative potential of disorder, so giving new insights into the process of change. It shifts the focus away from rational design and regulation to spontaneously self-organising processes.

Might notions such as these have implications for how we think about groups and organizations? If they did make sense in human terms they would imply a shift in thinking about change in terms of regulation at a boundary and design to thinking rather more in terms of self-organisation, the creative potential of disorder and emergent change. The focus would be on how a creative self-organising capacity might be promoted. How might we conceive of this in a group or organisation?

Other developments, such as the mathematical modelling of dynamical systems, raise questions about the very nature of boundaries in complex systems. Mathematical models of simple, iterative nonlinear systems (see, e.g., Stewart 1989) indicate that when energy or information flows through such systems at a low level, the system produces stable patterns of behaviour. When these flow at very high levels, the system displays unstable patterns of behaviour—in a sense it disintegrates. However, between these two possibilities, at some critical point in energy/information flow through the system, it displays a paradoxical dynamic that is both stable and unstable at the same time—it is mathematically chaotic or fractal. This paradoxical dynamic is a kind of boundary between stability and system disintegration, but it is a fractal boundary, that is, one in which self-similar patterns of combined stability and instability are found no matter how finely one looks for a dividing line, or region, between them. Might this have some implication for how we think about human interaction? How would we think differently if we move from a membrane metaphor of an organisation's boundary to a fractal metaphor in which it is problematic to say what is inside and what is outside?

Another strand in the field of nonlinear dynamics is the study of complex adaptive systems; a major interest driving many scientists in this area is understanding how life-forms evolve. A complex adaptive system is a network consisting of a large number of agents interacting with each other. Examples of complex adaptive systems are: the human genome consisting of 100,000 genes interacting with each other; the human brain consisting of 10 billion neurons interacting with each other; colonies of ants; traffic in a large urban centre; and an ecology of species.

Key questions are these: How do such complex nonlinear networks with their vast numbers of interacting agents function coherently to produce orderly patterns of behaviour? How do such living systems evolve to produce new orderly patterns of behaviour? An answer seems to lie in the process of self-organisation, that is, agents interacting locally according to their own principles, or intentions, in the absence of an overall blueprint, without any agent being in overall control of the system. When complex, self-organising networks are simulated on computers they display orderly patterns of behaviour that emerge from such self-organisation in what looks like the leaderless mess of many agents interacting with each other in the absence of an overall blueprint. Could this be true of human groups and organisations too?

The computer simulations of self-organising networks also display the three broad kinds of different dynamic referred to above: stability, disintegration or chaos, and between them, a paradoxical bounded instability at the edge of chaos. Furthermore, the conditions determining the kinds of dynamic the system displays are themselves characteristics of the network. Whether the network is stable, chaotic, or at the edge of chaos depends upon how responsive agents are in relation to each other; how richly connected they are to each other; and how diverse they are in relation to each other. In other words, the dynamics are determined by the pattern and nature of agent relationships and not by the nature of the agents themselves or by anything that is outside the network they form. Events outside the network in question may well have a substantial impact through its connections with those other networks that constitute its environment, but the nature of the response to this impact is primarily determined by the network's own internal dynamic—the principle of self-reference.

A conjecture of potentially major importance is that it is only when networks operate in the phase transition—bounded instability—at the "edge of chaos" that they are capable of evolving, that is, producing new patterns of relationship. When the dynamics are stable the network simply repeats its past: its own internal dynamics make it incapable of evolving novel responses to changes in its environment. A network that is "stuck" in this way would be endangered by the evolution of other networks it might be interacting with, eventually leading to its extinction. Similarly, operation in the chaotic or disintegrative dynamic would lead to extinction. At the edge of chaos, however, a network is capable of endless variety and thus novel responses to changes in its environment. A collection of networks in this dynamic follows a power law with many small and few large extinctions occurring while new species of network appear. The "edge of chaos" is thus not a guarantee of success for any agent or category of agents, but it is the dynamic within which the larger living network they are a part of evolves as new categories of agents emerge and old ones die. Networks at the edge of chaos evolve, that is, they learn from their own experience and they are thus history dependent. Does this notion of an intermediate dynamic illuminate creative human behaviour in a different way?

Finally, when a network functions in the paradoxical dynamic at the edge of chaos its long-term future is radically unpredictable: it operates in a state of uncertainty, evolving into an adjacent evolutionary space. Does this indicate anything about human groups and organisations?

THE NATURE OF HUMAN INTERACTION

If one is to employ the above constructs in understanding human interaction, then some kind of insight into human behaviour must be integrated into them. Both of the stories this chapter is concerned with do this by using certain psychoanalytical theories. What I want to do here, at the risk of oversimplifying both stories, is to explore how this integration is accomplished.

Miller and Rice (1967) used Bion's (1967) insights to see a group of people as an open system in which individuals, also seen as open systems, interact with each other at two levels. At one level they contribute to its purpose, so constituting a sophisticated (work) group, and at the other level they develop feelings and attitudes about each other, the group, and its environment, so constituting a more primitive (basic assumption) group. Both of these modes of relating are operative at the same time: when the basic-assumption mode takes the form of a background emotional atmosphere it may well support the work of the group, but when it predominates it is destructive of the group's work. In terms of the Miller and Rice model, one thinks of individuals as open systems relating to each across their individual boundary regions, so constituting a group which is also thought of as an open system with a permeable boundary region. Furthermore, they argue that it is confusing to think of organisations, or enterprises, as open systems consisting of individuals and groupings of individuals. So, an intersystemic perspective is adopted in which an enterprise is thought of as one open system interacting with individuals and groupings of them as other open systems.

Enterprises are seen as task systems—they have primary tasks that they must perform if they are to survive. Distinctions have been drawn between the primary task that ought to be performed, the primary task people believe they are carrying out, and the primary task they might be engaged in without being aware of it, the latter normally being a defensive mechanism. The primary task requires people to take up roles and, in order for it to be carried out, the enterprise, or task system, imports these roles across its boundary with the system consisting of individuals and groupings of them. Roles and relationship between roles fall within the boundary of the task system, but groups and individuals, with their personal relationships, personal power plays, and human needs not derived from the task system's primary task, fall outside it; they constitute part of the task-system's environment. So, we have one system, a task system, interacting with other systems, individuals and groups, and the latter are always operating in two modes at the same time: work mode and basic-assumption mode.

When the individual/group system has the characteristics of a sophisticated group with basic-assumption behaviour as a supportive background atmosphere, then it is exporting functional roles to the task system and the latter can perform its primary task. The enterprise, or task system, is thus displaying the dynamics of stability—that is, equilibrium or quasi-equilibrium. When, however, the individual/group system is flooded with basic-assumption behaviour it exports that behaviour into the task system so disrupting the performance of primary task. The dynamics are then those of instability or disintegration. Part of the enterprise as task system, a subsystem of it, might be set up to contain imported basic-assumption behaviour such as fight—its primary task is then to operate as an organisational defence that allows the rest of the task system to carry out its primary task. Without such organisational defences, the task system as a whole would import fantasies and behaviours that are destructive of the primary task—the dynamics of instability. These undesirable imports are to be diminished by: clarity of task; clearly defined roles and authority relationships between them, all related to task; appropriate leadership regulation at the boundary of the task system; procedures and structures that form social defences against anxiety; and high levels of individual maturity and autonomy. Most of these factors seem to me to emphasise design and global intention in the sense of some joint intention relating to the system as a whole. Furthermore, there is, it seems to me, a strong implication that the dynamics of stability are a prerequisite for a functioning task system, while the dynamics of instability are inimical to that functioning. There is little sense in this formulation of the creative potential of disorder.

Now consider how one might incorporate the same psychoanalytic insights from a complexity perspective. From this perspective boundaries are problematic so that one would not think in intersystemic terms with roles and task-defined relationships between them inside one system and groups, individuals and personal relationships in another, but rather, in terms of agents taking up task-related roles as well as many others, all of which are inseparable from each other. The focus then shifts from imports and exports across boundaries between enterprise system and group/individual system to a focus on the internal dynamics of enterprises consisting of groups and individuals, the dynamics being determined by the nature of their relationships. This is an important point of difference between a Tavistock approach and the complexity perspective I am suggesting. Miller (1993, p. 19) argues that the intersystemic view outlined above encourages us to focus on interdependence: people supplying roles to enterprises and those enterprises requiring performance in role from people in order to survive. He argues that when individuals and groups are seen as parts of the whole enterprise the focus shifts to a subordinate-superior relationship. However, from a complexity perspective one would think of people interacting, each affecting others in a circular fashion without any necessarily being superior or subordinate. From this perspective, individuals as agents are not exporting roles, and/or disruptive behaviour, to another system called an enterprise. Rather, the individuals as agents are interacting with each other at their own local levels, in certain respects in a self-organising manner, from which the nature of the enterprise emerges and this in turn affects the behaviour of the agents—cocreating each other. It is the nature of the relationships between individuals and groups that determines the possibility of emergent novelty, a point I will return to in the next section. In what follows, therefore, I will be talking about organisations and tasks emerging from the interactions between individuals and groups rather than task and individual/group systems interacting with each other in order to perform tasks.

Some of the key factors in the nature of the relationships between individuals and groups have already been mentioned: as agent responsiveness, connectivity, and diversity increase the group/organisation dynamic shifts from stability to bounded instability at the edge of chaos and then on into a disintegrative dynamic. Two further key factors, specifically pertaining to human relationships, must now be added—the need to take account of the second of these became clear to me from my general experience of group relations training conferences.

The first of these factors relates to the exercise of power, either in the form of authority taken from task or hierarchical structure, or in purely personal forms having little to do with such authority. If such power is exercised in a punitive, dictatorial, or very rigid manner it provokes either:

submission and conformity, in which case the system displays stable dynamics; or rage, rebellion, and sabotage, in which case it moves to the dynamics of disintegration. Both stability with its repetition of the past and disintegration are destructive of creativity. Erratic and arbitrary employment of power could generate similar dynamical possibilities, while the absence of power tends to provoke extreme rivalry, a disintegrative dynamic. Power relationships producing stable dynamics could be thought of in terms of basic-assumption dependency/pairing behaviour, while those producing disintegrative dynamics might be thought of in terms of basic-assumption fight–flight behaviour.

The second factor relevant to human dynamics is the level of anxiety experienced in relationships: the impact on the dynamic might be along the following lines. Stable dynamics are clearly evident when: relationships are such that there are clearly understood tasks with related roles clearly defined; leaders act in unambiguous, task-related ways; and organisational structures and processes provide adequate defence against anxiety and emotional factors are in the background. This is, of course, Bion's sophisticated work group. However, stable dynamics could also be characteristic of relationships primarily constituted to avoid anxiety, such as excessively rigid procedures or basic assumption dependency behaviour. Disintegrative dynamics would be evident when, for whatever reason, high levels of anxiety are responded to in the form of basic-assumption fight–flight behaviour.

What I am suggesting then is this. Stable group/organisational dynamics occur either when a sophisticated work group is evident, that is, when people are carrying out tasks they are clear about, in relational conditions conducive to doing so; or in conditions in which they are avoiding anxiety by operating as a basic-assumption group. Disintegrative dynamics would occur when people in groups/organisations are suffused with anxiety, once again operating as a basic-assumption group. Neither of these dynamics, that of the work group or of the basic-assumption group, really enables an understanding of how the novel or the creative comes about. In novel situations people are struggling with finding out what the tasks might be and so cannot yet be a work group in the terms described above, but since people often succeed in producing something novel they cannot be a basic-assumption group either. I am arguing that the formal Tavistock model, with its intersystemic formulation, its emphasis on clarity of primary task, and its distinction between work and basic-assumption group has difficulty in accommodating the whole question of creativity. I am also suggesting that a complexity perspective offers the potential for a more integrated approach to this question. I am claiming that it is this that a complexity perspective might add to the Tavistock model.

The developments in nonlinear dynamics outlined above suggest looking for a phase transition—bounded instability—between the dynamics of stability and disintegration, between the work group (with basic-assumption behaviour in the background) and the much more prominent basic-assumption group. What I am suggesting here is the need for a paradoxical concept of dynamics that is more complex than Bion's two modes of work and basic assumption simultaneously present. Does experience in organisations also indicate the need for this and if so what would be the human dynamics of bounded instability?

It seems to me that the notion of a group/organisation operating in a state of bounded instability, between a work group (with basic-assumption behaviour in the background) and a more prominent basic-assumption group, is needed to make sense of many situations managers in organisations find themselves in. When they are concerned with reasonably repetitive day-to-day activities, or are developing their activities within reasonably clear existing strategic directions, they do display the characteristics of the work group (with basic-assumption behaviour in the background) given above: they are reasonably clear on their shared primary purpose; they operate with authority derived from their tasks and their position in the hierarchy; the importance of wider personal relationships is not that great. At other times when their leaders are exercising their power in arbitrary and erratic ways, when they face extreme uncertainty in situations where trust in personal relationships is largely absent, the operation of a much more prominently basic-assumption activity comes to the fore. But in my experience there are often situations in which something between these possibilities occurs. I am thinking of situations of considerable uncertainty in which it is not clear what the primary task means— even in organisations such as educational establishments that seem to have always had the same primary task, its meaning changes over time. This seems to me to be even more prevalent in commercial enterprises where managers have conflicting tasks and find it impossible to agree for long on which of these is primary. In such situations managers are working in a state of not knowing what the task is and their roles emerge as they struggle to identify tasks or develop their meanings in a novel way. Such situations are anxiety provoking, but in my experience this does not always lead to predominating basic-assumption behaviour. What I observe in these situations is managers engaged in intensive conversations with each other outside any formal forum in which they demonstrate an ability to hold the level of anxiety between them. This possibility seems to be rooted in the quality of their personal relationships. Conversations of this kind proceed in a self-organising manner without anyone knowing the direction they will take. Leadership shifts according to who has a contribution to make and this often does not coincide with authority derived from existing tasks or hierarchical position. Personal power plays may be evident but are exercised with some coherence and responsibility consistent with a good-enough level of trust in personal relationships.

What I am describing is the operation of a community of practice embedded in personal relationships creating social and also covert political networks in which learning may take place and innovations may emerge, despite the lack of clarity and coherence that would be needed for it to qualify as a work group in Bion's terms. The psychodynamic forces at play in such groups certainly include aspects of basic-assumption behaviour, but this concept is not sufficient to understand what is happening. What else is necessary?

Nitsun (1996) points to the extreme positions on group life taken by Bion (1961) and Foulkes (1964). Bion's view of groups is an essentially negative one in which individuals are at war with their groupishness and are continually in danger of being overwhelmed by primitive group processes. Foulkes, on the other hand, idealises groups and their potential for learning and behavioural change, seeing a high probability of changes in individuals emerging from their relating in a group. What might be necessary in understanding the kind of situations I outline above, therefore, is the addition of some of Foulkes's optimism about group behaviour to Bion's pessimism.

Foulkes focused on the network, or matrix of verbal and nonverbal communications that emerge as a group comes together and develops relational patterns, where that matrix in turn affects individual functioning. It is a short step from the position adopted by Foulkes to social constructionism in which the ordinary everyday reality we all act into is seen as being constructed by the ordinary everyday conversations we engage in (Shotter 1993). Alongside this emphasis on the constructing nature of conversation there is also the importance of nonverbal communication (Wright 1991, Meares 1992) and prereflective processes (Stolorow, Atwood, and Brandchaft 1994) that continue throughout life. In addition to the splitting and projective processes emphasised by Bion, Foulkes pointed to the interpersonal processes that create and transform, such as mirroring, affirmation, and resonance (Foulkes 1964). Bollas (1987, 1993) talks about the transformational potential of processes of giving-receiving and evocation in which individuals are transformed by relationships. His discussion of the "unthought known" is relevant here—he is talking about the intuitive and imaginative process in which people struggle to articulate what they know at a prereflective level into verbal communications. These notions direct attention to what people in groups and organizations do when they do not know what they are doing in relationships that enable them to hold the anxiety provoked as the prelude to the potential emergence of something creative. Winnicott's (1965) notion of transitional phenomena is also relevant here: we might think of a good-enough holding environment as enabling groups of people to operate as a transitional group, a group in a state of not knowing what its task is yet, a group that must therefore employ creative imagination, fantasy, and play (Miller 1983).

What I am suggesting, then, is this: at some critical point in information flow, individual connectivity and diversity, responsible exercise of power, and relational ability to contain anxiety, the dynamics of a group/organisation shift from those of the stable work group to a dynamic at the edge of the disintegrative basic-assumption group. While the work group is characterised by design and shared intention and the basic-assumption group by self-organising functioning of a disintegrative kind, a group at the edge, in bounded instability, displays both fantasising of a basic-assumption kind and also the kind of transformational processes that Foulkes, Bollas, and others allude to. Such a group is then in a state of not knowing and it self-organises with the potential for the emergence of the novel.

A COMPLEXITY PERSPECTIVE ON ORGANIZATIONS

Turning now specifically to the level of organisations, both a Tavistock and a complexity approach focus attention on interaction. In the former, an organisation consists of two systems with a task system importing roles across a permeable boundary from another system consisting of individuals and groups. In the latter, individuals, in their many roles and relationships, are seen as agents in an organisational network of relationships. The difference is, once again, that a Tavistock approach focuses our attention on regulation at the organisational boundary while a complexity perspective directs attention to the self-organising nature of the internal dynamic, that is, the nature of the personal relationships between the agents. Before looking at the consequences of this difference in focus we need to consider whether there is any justification in our experience of life in organisations for thinking that there are any self-organising processes at play. To argue that there are, I suggest drawing a conceptual distinction between two intertwined aspects of an organisation's network of relationships, even though in practice they cannot be separated (Stacey 1996a,b).

I give the label legitimate relationships to the first of these aspects. These are the more visible aspects of any organisation, and they embody organisational members' shared interpretation of their principal tasks. These relationships take the form of a designed set of roles required to carry out the tasks, organised into a hierarchy establishing formal authority. They are governed by a designed set of procedures and rules comprising a bureaucracy and also by defined habits and routines accumulated from past task-performance, as well as the widely accepted, officially approved, shared ideology at any given time. This set of legitimate relationships is a blueprint that has largely been designed and installed at some point and its purpose is to enable stable joint action in carrying out the principal tasks of an organization as they are understood at the time. But how did this organization reach the point where it had enough shared sense of task and role and could therefore install some designed set of legitimate relationships to carry it out?

We all know that the process leading up to changes in the legitimate aspects of an organization is a messy, confusing, and often disturbing one that threatens vested interests, that may well provoke covert politics, basic-assumption behaviour, and the like. But in my experience this is not all that happens, and if it were it would be hard to explain how anyone ever gets to the stage where there is enough agreement and clarity to embark on anything that qualifies as rational analysis and design. In the behaviour that leads up to changes in legitimate relationship and reformulations of what the task is, I experience a process of learning occurring through conversations taking place in the context of personal relationships. The experience is one of dialogue that slips in and out of debate, in which people are trying to surface what they do not yet know, as well as articulating what they know but cannot yet express. It is through conversing, persuading, and exerting influence in a highly personal network of relationships that the shape of potential change emerges. Only then can it appear as rather more orderly, rational work. This network of personal relationships, with all their social and psychodynamic implications, constitutes what I think of as "shadow relationships" that underlie and intertwine with the legitimate ones. One would not think in terms of boundaries between legitimate and shadow relationships since everyone is always simultaneously engaged in both.

Now these shadow relationships are quite clearly spontaneously self-organizing. Joining an organization means taking up designed roles in a network of legitimate relationships, closely related to currently identified tasks. But as soon as we do this, at the same time, we spin a network of personal relationships—social, emotional, psychological, and political—that are as vital to our work as is the role we take up in the legitimate network. No one tells us who to network with, and we cannot individually determine what that network will be because it requires the acceptance and cooperation of those we network with. These self-organizing social/psychodynamic/political relationships can and do produce corruption, vicious personal striving, covert politics of a harmful kind, basic-assumption behaviour, and sometimes even near-psychotic fantasies. But they can also function as the basis of a learning community of practice, the location of an organization's narrative and tacit knowledge, the vehicle for the exploration of intuitions and, as such, the process of organizational learning and the origin of an organization's creativity. What emerges out of the interaction of agents in the shadow network may ultimately come to be embodied in changes in its legitimate pattern of relationships. No one fully understands their organization's network of shadow relationships, no one is in control of it, but all contribute to what it is by interacting in their own local network and by so doing they may learn and so change as individuals as well as producing emergent change in their organization as a whole. The shadow relationships seem to me to be characterised by the same kind of coevolutionary process that we see in all other complex processes.

I suggest it is possible to see the three kinds of dynamic we encounter in complex systems at work in organizational shadow relationships. We might think of shadow relationships being characterised by basic-assumption behaviour, and this may be either disintegrative of work and learning or trap members in stable repetitive fantasies, such as the fantasy that they are a family. On the other hand, shadow relationships may embody a basic assumption that is supportive of the functioning of the work group. We can also see rigid social defences (Jaques 1955, Menzies Lyth 1975) producing highly stable dynamics, but at the price of interfering with learning and a halt to an organization's evolution, or on the other hand, as enabling it through enough anxiety containment to operate as a work group. But there is another possibility, namely, the one described above in which people use their close personal relationships in a manner that qualifies, much of the time anyway, as some kind of learning process in which they work with what they do not yet know. I think these are the dynamics of the phase transition at the edge of disintegration for an organisation, and I say this because people learning in the way I have described are in fact subverting or undermining existing legitimate relationships in the interest of changing them so that they survive—a paradoxical form of behaviour, a mixture of stability and instability.

There is one other point I want to make about shadow relationships that relates to its boundaries. The boundaries around a set of legitimate relationships are clear—we know who is a member of an organisation and who is not. But the boundaries around shadow relationships are far from clear—we have to think of such boundaries as irregular and fractal because these webs of personal relationships stretch across into other organisations and institutions and are used to carry out the work of the organisation.

It may be helpful at this point to draw some comparisons between the legitimate/shadow distinction and other similar distinctions. A well-known distinction has been made for a long time between formal and informal systems in organisations. This does not coincide between the legitimate/shadow distinction. First it has become clear to me that the legitimate and the shadow distinction relates not to different systems but to different kinds of relationship. Legitimate relationships include more than those that are formally defined in terms of hierarchies and bureaucracies—they also include those conforming with the shared ideology of an organisation, normally regarded as part of the informal system. Turning to the Tavistock model, the task system/work group consists of roles rationally deduced from the primary task or unconsciously assumed as defences against anxiety that enable task performance. It is largely a designed system and is, in a sense, an ahistorical view—it is derived from current tasks rather than the emergence of new legitimate relationships from changing shadow ones. We introduce historicity when we postulate that the legitimate pattern of relationships at any one time is that which has been articulated/designed as it emerged from previous interactions in shadow relationships. The requirement for novelty to emerge is shadow relationships that constructively subvert existing legitmate ones—this notion of the connection between subversion/destruction and creativity seems to be absent in the work group/basic-assumption group distinction.

I suggest that this distinction between simultaneously present legitimate and shadow relationships highlights the different emphases between

Tavistock and complexity perspectives. As I see it, a Tavistock approach directs our attention to legitimate relationships and the dangers posed to them when the dynamics of shadow relationships are those of disintegration—the dynamics of the basic-assumption group and corrupt exercise of personal power. The self-organising nature of the basic-assumption group and its destructive effect on the ability to think is powerfully experienced at group relations training conferences. An organisation's propensity to engage in this dynamic can damage its performance of current tasks and also significantly affect its strategic direction, either locking it in to its existing direction or producing emergent strategies based more on fantasy than anything else (Miller 1977).

By contrast, a complexity perspective directs attention to organisational life in a somewhat different way. It focuses attention on both the negative and the positive aspects of shadow relationships and their potential for self-organising learning and the emergence of innovation, as well as their propensity to attack such learning. This might suggest looking for the means of engaging with and so encouraging the positive aspects and containing the negative aspects of shadow relationships itself, without surfacing them much into the legitimate sphere.

PRACTICAL IMPLICATIONS FOR CONSULTING

In order to explore some practical implications of the different perspectives discussed above I am going to review a consultation (Shapiro and Carr 1991) in which the consultant utilised a Tavistock perspective. There are, of course, variations from consultant to consultant in interpreting what a Tavistock perspective might mean in practise, just as there are as many views about what complexity means as there are people writing about it. The description of the consultant's role is summarised below.

For Shapiro and Carr the consultant uses countertransference feelings to formulate hypotheses about the transferential and projective processes at work in an organization, and about the impact of basic-assumption behaviour on the work of that organization. They see the function of the consultant as one of feeding back those hypotheses into the life of the organization and so fostering a collaborative, negotiated understanding and verbalization of the unconscious, irrational processes at play. It is believed that this process enables the reclaiming of projections and distorted impressions of reality, so restoring to the group its work function. The consultants engage with and understand the complexity of organizational life by adopting an interpretive stance. This stance is seen as the most important element in creating a holding environment and they draw an analogy with a therapeutic setting: containment and holding ordinarily refer to symbolic interpretive ways in which the therapist manages the patient's (and his or her own) feelings (Shapiro and Carr 1991, p. 112). Another feature of the holding environment, one that interpretation aims to secure, is the clarity of task, boundary, and role. This is seen as containing, for example, sexual and aggressive feelings. Empathic interpretation affirms individuals in their roles and the resulting containment establishes a holding environment which provides for safe regression, a shift from rationally organized words to the primitive distortions of fantasy images and simple metaphors which can then be articulated and so disarmed. The aim of interpretation is to move people from states of irrational anxiety and fantasy that distort work to more reality-based taking of roles that support it.

A complexity perspective puts a different emphasis on the consultant's role and activity. It tends to de-emphasize interpretation and ascribe much more importance to the quality of personal relationships. More than interpreting or formulating and presenting hypotheses, the consultant facilitates participation in the shadow activity of an organisation. The consultant's role here would have some similarity with Bollas's (1987, 1993) view of the role of the therapist. He attaches importance to patients developing a capacity for self-analysis. He sees this as requiring tranquility, creating the conditions for the arrival of insight through reception and evocation. This means avoiding interpretive activity and waiting patiently for the patient to evoke what is within him or her. Through this process he sees images and ideas emerging where they had not existed before. The therapist is then not a container but an auxiliary in the evocation of new inner experience. Bollas is talking here about tapping a form of knowledge that permeates a person's being but is not yet known. The analyst assists in the process of evocation by being there as a presence, strictly speaking, neither inside nor outside the analysand's mind.

From a technical point of view, then, the theoretical shift implied by a complexity perspective would be accompanied by a practical shift from the emphasis on interpretation and containment by task, boundary, and role clarity. It would shift to an emphasis on the consultant as a facilitator of the organizational evocation process in which groups work within an organization's unthought known, the nonverbal, prereflective knowledge embodied in its patterns of relationship. Holding is then provided by these patterns of relationships that the consultant contributes to by being there—it is the group matrix that is the container, not just the consultant. Perhaps the most important aspect of the consultant's being there is the way he works and the practices and behaviours that this helps to affirm. The emphasis shifts away from concern with the primary task, boundary, and roles, to how, through patterns of personal relationship, people may cope with the dynamics at the edge of disintegration where they struggle to give expression to what they might know but cannot yet think.

I can perhaps best illustrate this shift by taking a case that Shapiro and Carr (1991) describe. The case concerns a consultation to the Adolescent and Family Treatment and Study Center within the McLean Hospital in Massachusetts. The organizational problem had to do with staff members feeling isolated from one another and finding it difficult to find a way of collaborating to further develop the Center's treatment, training, and research. The presence of the consultant seemed to free people to talk in unfamiliar ways, even though, as they note, he had not yet done anything. Furthermore, in his conversations the consultant received many personal confidences as if those working with disturbed families and adolescents could find no way in their organizational life to share their own disturbances in personal and family life. This was despite their reporting that no one was able to say or do anything without being analysed because every person's behaviour was assumed to provide some evidence relating to the clinical treatment of patients. The one person who could be relied upon not to interpret was the program secretary who then became the repository of personal confidences. In addition to the secretary's desk at reception there was another interpretation-free zone in the nurse's station. In both places relaxed and interpretation-free conversation amongst the staff took place.

However, because of the all-inclusive theory of the treatment program, the consultant regarded these zones as intrinsically illegitimate—even subversive—and in doing so he might have been reflecting what the members of this organization themselves thought. The consultant concluded that these informal subgroups existed because people felt unsure about their roles and were therefore having difficulty finding appropriate means for sharing feelings about their roles. The informal subgroups were seen as a compensation for some lack in the formal system and the consultant saw this as a managerial obstacle to the theory the entire staff accepted about their way of working. This led them to think through the roles of patients in their families and their own roles in relation to the Center as a family parallelling the families the Center was dealing with. The informal groups were thought to have become "waste containers" in that important affective communication was simply discharged instead of being examined. He thought that there was a struggle between staff in their formal roles and in their roles as individuals because they were being instructed to share experiences rather than negotiating such sharing. In addition, the consultant identified a lack of organizational clarity in the relationship between different units that made up the Center. He furthermore stressed unclear chains of authorization for people to take up their roles.

The consultant concluded that the organization was not functioning according to its design, for all the reasons set out above, and therefore considered how tasks could be understood by all so that roles within and across units could be legitimately authorized and fully integrated. The consultant recommended clarification of authorization from one level to another in the hierarchy and established structured meetings, a policy council, and a management team in order to promote effective communication. To deal with the feelings of the staff in a legitimate manner, the consultant stressed the need to develop a culture, with the authorization of the director of the Center, in which interpretation was to be encouraged, but only within limits relevant to the work. People would then bring their feelings to legitimate forums where they could be made available for examination in relation to the work rather than discharged in informal subgroups; such examination would constitute an interpretative stance, that is, a collaborative verbalization of unconscious processes leading to withdrawal of projections that might be adversely affecting task performance. The objection to the subgroups thus seems to be based on the belief that, since they are based purely on personal relationships rather than task, they are fertile ground for projections and basic-assumption behaviour. Later, an invitation by a section head to the director to take a consulting role in her unit was seen as evidence of the internalization of the interpretative stance. However, as the organization experienced further change it was noticed that the secretary became even more active in her informal role.

Note how the consultant's model of organizational functioning leads him to focus on the network of legitimate relationships in the organization—he recommends the clarification of formal tasks, the authorization of formal roles, and the authorization of culture and the removal of informal subgrouping through more formal participation, that is, through adopting an open, collaborative, and public-interpretative stance. He notes that people seem to develop new ways of talking informally, but attaches little importance to it. The significance of the increasing activity of the secretary is not explored.

By contrast, a consultant working from the kind of complexity perspective outlined above would make sense of the situation at the Center in a different way and would, therefore, operate in a different manner. The manifestations of this organization's network of shadow relationships, the interpretation-free zones would not be regarded as evidence of organizational malfunctioning but potentially, anyway, an essential part of its flexibility and the ultimate source of its creativity and ability to change. So, although the consultant would formally contract with the legitimate aspect of an organization, he or she would contract to work primarily in the network of shadow relationships by in effect joining and taking part in them in order to attempt to understand their dynamics. However, this understanding would not normally be used to interpret, since the approach is not primarily educative but participative. Through participation the consultant would be affirming the usefulness of informal subgroup activity based on personal relationships and seeking to seed further activity of this kind (Shaw 1997).

One of the main purposes in joining the shadow networks in this way would be to foster new ways of conversing, since it is in ordinary everyday conversation that we construct organizational realities, and as we saw in the above case this is often done simply by being there. This view is in keeping with evidence on how, far from being a distraction, personal, relationship-based subgroups constitute communities of practice and in their informal conversations embody the most up-to-date knowledge that create the frontiers of organizational learning (Brown and Duguid 1991). The complexity perspective I am suggesting sees shadow relationships as essential to organizational flexibility and creativity, the matrix within which the unthought known might come to be thought, the activity that might be saying something important about organizational malfunctioning and be the seedbed for changing it. For example, the shadow activity at the secretary's desk might be suggesting that an element of the officially accepted way of working in the Center is actually not accepted, perhaps because it is not all that appropriate. The consultant might then take part in conversations about whether it is appropriate for professionals in a Center providing services to families to think about that Center as a family and themselves as members of it. After all, organizations are not families other than in fantasy. When one works with shadow relationships in the manner I am suggesting, the question of authorization becomes unimportant. Shadow relationships are self-organizing and people spontaneously take up their shadow roles on the basis of unspoken mutual consent and usefulness.

Does this approach to consulting mean that the consultant does not work with the unconscious? I do not think so. The consultant would still notice projective activity when it was apparent and also use his or her countertransference feelings. I suspect, though, that there would not be an a priori expectation of strong transferences or projections. The consultant working from the perspective I am describing would also be aware of other forms of psychic-relating such as the mirroring, resonance, and evocation mentioned earlier on. However, the difference is that such feelings are part of a total participation in the shadow network and not used primarily for interpretive purposes.

CONCLUSION

This chapter has sought to view a Tavistock approach from a complexity perspective. It has suggested how the latter perspective might complement, and contrast with, a Tavistock approach. A complexity perspective suggests a shift in focus. The emphasis in a Tavistock approach is placed on design and regulation at the boundary between task and individual/group systems. This tends to privilege what I have called the legitimate aspects or relationships in organizations. A complexity perspective does not make the distinction between task and individual/group systems but, rather, considers the impact of relationships between individuals and groups on the dynamics of an organisation seen as one network nesting in larger networks, such as markets. The emphasis is then placed on the self-organising, emergent properties of the shadow aspects of relationships in organizations, on how they coexist with the legitimate aspects and tend to be subversive of those legitimate aspects in the interests of change. It has been argued that this shift in focus directs attention to the potential for novelty and creativity.

REFERENCES

Bion, W. R. (1961). *Experiences in Groups and Other Papers*. London: Tavistock Publications.
Bollas, C. (1987). *The Shadow of the Object*. London: Free Association Press.

————— (1993). *On Being a Character*. London: Routledge.

Brown, J. S., and Duguid, P. (1991). Organizational learning and communities of practice: toward a unified view of working, learning and innovation. *Organisational Science* 2:40–57.

Emery, F. E., and Trist, E. L. (1960). Socio-technical systems. In *Management Sciences, Models and Techniques*, vol 2, ed. C. W. Churchman and M. Verhulst, pp. 83–89. Oxford: Pergamon.

Foulkes, S. H. (1964). *Therapeutic Group Analysis*. London: George Allen & Unwin.

Gell-Mann, M. (1994). *The Quark and the Jaguar*. New York: Freeman & Co.

Gleick, J. (1988). *Chaos: The Making of a New Science*. London: William Heinemann Limited.

Goodwin, B. (1994). *How the Leopard Changed Its Spots*. London: Weidenfeld and Nicholson.

Jaques, E. (1955). Social systems as a defence against persecutory and defensive anxiety. In *New Directions in Psychoanalysis*, ed. M. Klein, P. Heimann, and P. Money-Kyrle. London: Tavistock Publications.

Kauffman, S. A. (1993). *Origins of Order: Self Organization and Selection in Evolution*. Oxford: Oxford University Press.

————— (1995). *At Home in the Universe*. New York: Oxford University Press.

Levy, S. (1992). *Artificial Life*. New York: First Vintage Books.

Meares, R. (1992). *The Metaphor of Play*. Melbourne: Hill of Content.

Menzies Lyth, I. (1975). A case study in the functioning of social systems as a defence against anxiety. In *Group Relations Reader*, ed. A. Coleman and W. H. Bexton. Sausalito, CA: GREX.

Miller, E. J. (1977). Organisational development and industrial democracy: a current case-study. In *Organisational Development in the UK and USA: A Joint Evaluation*, ed. C. Cooper, pp. 31–63. London: MacMillan.

————— (1983). Work and creativity. *Occasional Paper No. 6*. London: The Tavistock Institute of Human Relations.

————— (1993). *From Dependency to Autonomy*. London: Free Association Books.

Miller, E. J., and Rice, A. K. (1967). *Systems of Organization: The Control of Task and Sentient Boundaries*. London: Tavistock Publications.

Nitsun, M. (1996). *The Anti-Group: Destructive Forces in the Group and Their Creative Potential*. London: Routledge.

Prigogine, I. (1996). *The End of Certainty*. New York: The Free Press.

Prigogine, I., and Stengers, I. (1984). *Order Out of Chaos: Man's New Dialogue with Nature*. New York: Bantam.

Shapiro, E. R., and Carr, W. A. (1991). *Lost in Familiar Places*. New Haven, CT: Yale University Press.

Shaw, P. (1997). Intervening in shadow systems of organisation: consulting from a complexity perspective. *Journal of Organizational Change Management* 10:235–250.

Shotter, J. (1993). *Conversational Realities*. Thousand Oaks, CA: Sage.

Stacey, R. D. (1990). *Dynamic Strategic Management for the 1990s*. London: Kogan Page.

—————— (1996a). *Strategic Management and Organisational Dynamics*. London: Pitman.

—————— (1996b). *Complexity and Creativity in Organizations*. San Francisco: Berrett-Koehler.

Stewart, I. (1989). *Does God Play Dice?* Oxford: Blackwell.

Stolorow, R., Atwood, G., and Brandchaft, B. (1994). *The Intersubjective Perspective*. Northvale, NJ: Jason Aronson.

von Bertalanfy, L. (1968). *General Systems Theory: Foundations, Development and Application*. New York: George Braziller.

Waldorp, M. M. (1992). *Complexity: The Emerging Science at the Edge of Chaos*. Englewood Cliffs, NJ: Simon & Schuster.

Winnicott, D. W. (1965). *The Maturational Process and the Facilitating Environment*. London: Hogarth Press. Reprinted 1990 by Karnac (Books) Ltd.

Wright, K. (1991). *Vision and Separation*. Northvale, NJ: Jason Aronson.

5

Enemies Within and Without: Paranoia and Regression in Groups and Organizations

H. SHMUEL ERLICH

INTRODUCTION

I have chosen to use the opportunity occasioned by being asked to contribute to this volume to raise some fundamental issues regarding paranoia and enmity. I see this as tremendously significant to all of us—as people who share an interest in group processes and their manifold implications, but also as citizens of society and its institutions. I strongly believe that this issue—of paranoia and enmity—crucially affects our everyday living and survival.

PARANOIAGENESIS IN GROUPS AND INSTITUTIONS: AN OVERVIEW

Paranoiagenesis in groups and organizations, in institutions and in large social systems, is ubiquitous. As Israelis, we are acutely aware of the powerful, sometimes deadly, effects of group paranoid processes, particularly as they suffuse and endanger the current peace-making process. One way of addressing paranoiagenesis is to look at what an "enemy" is, and how we relate to him. I would, therefore, like to explore this issue from the combined vantage points of psychoanalysis and group relations theory and experience.

Recent events in Israel have demonstrated once again, if such demonstration were needed, the critical significance of what it means to be paranoid—not necessarily in the clinical sense, but in the group sense, at the organizational and social level. Time and again it happens that individuals serve causes and aims, typically political and power-seeking, which many of us, of different political leanings and affiliations, would designate as clearly paranoid. Yet, such individuals can almost never be defined clinically as paranoid or psychotic, precisely because their paranoia is contained and supported by a group that gives it consensual validation and political respectability of sorts. Such individuals are not the lonely, deranged, withdrawn, hallucinatory, clinically delusional, mentally ill persons met with in clinics and hospitals, or artfully studied in movies like *Taxi Driver*. Their paranoia is a much more difficult kind to identify and deal with. It takes refuge in commonly held opinions and well-known political positions, which are typically pushed to their most extreme limits and implications. To the outside observer it may appear that these positions can still be treated rationally, and that they can be logically and democratically debated. They always hinge, however, on an identification of "The Enemy" and on a specific prescription for dispensing with "Him." Like many readers, I see these people and groups as highly dangerous; but this view in no way detracts from my other, equally firm conviction, that their paranoia and enmity is in the service of all of us. To brand these persons and groups as lunatics, fanatics, or paranoid may entrap us in the same process that gave rise to them in the first place, no matter how accurate such descriptions might be. The paranoid label, in this sense, identifies "an enemy" who can then carry all sickness and badness in him, after we have evacuated these parts of ourselves and projected them into him. In order to maintain and continue this self-cleansing process, we need to uphold and maintain these projections, which hinge on viewing these persons as "sick" or "evil," the source of all our problems and badness. Projective processes of this sort allow us to feel relatively sane, safe, and good, and to sustain these feelings as long as we have an enemy on whom we can continue to project our unwanted parts. Accordingly, my focus in what follows will be on those aspects of group process that may be described as regressive, and especially on the development of paranoia and enmity within them. I will then reflect on the meaning and universal presence of enemies, and on how this is related to the experience of otherness and boundaries, as well as of fusion with the other.

Freud (1921) saw in the twin processes of identification and introjection the two primary dynamics operating in the formation of systems, or what he called "artificial groups," such as the army and the church. The early Kleinian psychoanalysts who expanded these notions and deepened our understanding of group processes have extended these insights, describing projective and introjective identification as the two primary mechanisms operating in groups and organizations. In a groundbreaking paper, Elliot Jaques (1955) described the ubiquity of paranoid and depressive anxiety as well as their containment by social defenses. As he put it, "All institutions are unconsciously used by their members as mechanisms of defense against . . . psychotic, paranoid and depressive anxiety." Conversely, their function as containers pulls individuals into them. The cooperation of individuals within institutions is motivated and shaped by their unconsciously shared fantasy picture of the aim, form, and content of the institution. Where the institution or organization no longer serve this unconscious function, the ground is ready either for social change or for regressive breakdown, which often takes the shape of paranoiagenesis (Kernberg 1994).

ORGANIZATIONAL PATHOLOGY

Group behavior and pathology have been conceptualized at several levels of abstraction. Looked at from the intrapersonal level, the group is understood as the sum total of the dynamics of the individuals making it up. Individual regression in groups is here explained by the operation of individual intrapsychic mechanisms summed up across individuals. I have previously referred to this as the level of "summation" (Erlich 1996, 2000).

At the next, interpersonal level, the emphasis is on the interaction between individual dynamics and the group framework. The guiding principle is that individual human development can reach its full culmination and ultimate perfection only within social groups and institutions. We may think of this as the level of "consummation" (Erlich 1996, 2000). In psychoanalysis, this level is best represented by the various schools of object relations.

Elliot Jaques's description of the containment of individual paranoid and depressive anxiety by social structures serves as a bridge to the next level. At this third, suprapersonal level, we meet with concepts of the "group-as-a-whole" or "group mind." Here the group is an emergent and organic entity in its own right, although its understanding still employs concepts borrowed from individual psychology. It is this level that Bion (1961) developed in his seminal work on groups. The group-as-a-whole is conceptualized as a maternal entity, evoking reactions and conflicts in its members which are reminiscent of those experienced by the infant towards his or her mother. Group life involves struggles around wishes for merger and fusion as against separateness and loneliness; powerful experiences of being fed and satiated, or being denied and frustrated; acute ambivalence and overwhelming tensions between engulfment and estrangement; and defensive recourse to splitting and projective identification. Participation in the group process evokes early, almost primary, anxiety and regression, typical of Melanie Klein's (1946) paranoid-schizoid position. Through the massive use of projective identification, fragments of the self are expelled and forcefully infiltrated into the object, while continuous contact and identification with them is constantly maintained. This accounts for the experience of individuals in the group who feel as if parts of themselves are invested externally, in the "mother-group" and the others who make it up. A covert, implicit, symbolic, and largely unconscious common network is woven, which gives rise to the fantasy of the group as a new and emergent entity. Out of this, and depending on the specific group needs and tasks, compartmentalization of functions is fashioned, individual group members are induced into specific roles, and the particular culture that characterizes the group takes shape (Wells 1985).

The fourth level of discourse is the systemic, intergroup, or organizational level. The focus here is primarily on systems and their component subsystems, as well as on intergroup and organizational relatedness and relationships.[1] From the vantage point of his or her role within systemic group relations, the individual is regarded as expressing and symbolizing the group he represents; and the manner in which he or she is treated represents relatedness to his or her reference group, and not to him or her as a person.

With these considerations in mind, let us now return to paranoiagenesis at the individual and group levels. We must immediately observe that in regard to paranoia there is but a thin line between the individual and the organizational paranoid dynamics. When speaking of paranoia, we are dealing with a phenomenon that, by its very nature, is at the interface of these two levels. The clinical point of view typically focuses on the individual aspects of the paranoid experience, mainly on the loss of portions of one's adequate

---

1. The terms "relatedness" and "relationship" in their present connotation were introduced within the Tavistock Institute's Leicester Conference (Miller 1989). This distinction usefully refers to the difference between actual "relationships," involving external interactions and transactions, and the more internal, fantasy-level elements of "relatedness," which may permeate and affect the former, for example: attribution, projection, and transference.

functioning. This point of view will outweigh and shift the focus away from the above-described third and fourth levels of discourse to the first and second ones, that is, to the intra- and interpersonal. It appears that both clinically and organizationally we prefer to deal with paranoid experience as though it is lodged in the individual, and to regard it within intra- and interpersonal levels of discourse. Not enough emphasis is placed on what to my mind are no less important and urgent questions: To what extent may we speak of groups, institutions, and organizations as deranged and paranoid entities? And what may be the causes of such regressions? What insights and understandings can we fall back on in order to understand phenomena like hatred for strangers, or paranoid rivalries between competing institutions? Under what circumstances are these exacerbated, and when does full-blown psychotic and paranoid institutional, group, or organizational behavior emerge? Can we diagnose the occurrence of organizational paranoia with certainty, or do we content ourselves with the insight that all human institutions are breeding grounds for paranoiagenesis? Is it indeed primarily a function of the individual psychopathology of the specific leader in power at the particular historical moment—the Hitlers or the Stalins? Under what circumstances are less pathologically inclined leaders—the Trumans, Bushes, and Clintons, the Gorbachevs and Yeltsins, to mention only a few—also capable of unleashing paranoid delusions? When does normal rivalry and competition between different schools of thought, whether of art or psychoanalysis, turn into regressive enmity and paranoid slander? Where is the point when the end starts to justify the means, and how can we become more aware of it? Can we draw a line between prejudice and paranoiagenesis? These are some of the questions in need of answers if we wish to improve our understanding of organizational and social paranoid phenomena. In sum, it is necessary to advance beyond the individual and interpersonal levels in order to understand more fully issues of group and social paranoia, of enmity and international paranoia.

THE NATURE AND PLACE OF PARANOIA

Paranoia consists in pathological distortions at the very heart of reality perception and testing, falling within and severely undermining the ego function of judgment. The capacity for evaluating and judging reality develops gradually out of the extraordinarily refined attunement of internal experiential states and internalized social and cultural standards. Social context and group culture clearly co-determine the ego function of judgment. Mental disturbance, and in particular paranoia, is often regarded as stemming from purely intrapsychic dynamics. This view, strongly propounded by psychoanalysis, is patently simpleminded. It is based on a unidimensional view of subjective experience. Socially shared psychic reality goes well beyond intrapsychic factors, and indeed contributes to them. Yet, to make the links between these levels is not as simple as we may wish, precisely because of the prevailing dichotomy between intrapsychic and group levels of discourse.

Nonetheless, individual "clinical" paranoia clearly owes a great deal to the social and institutional climate that one lives and works in. Regression, the loss of adequate judgment and reality assessment by the individual, is co-influenced and co-determined by factors that exist at the interpersonal/family/group level and at the institutional/organizational/cultural level. These may be directly or indirectly related to the individual's specific psychopathology—the individual's repressed homosexual desires, overt or covert aggression and grandiosity, narcissistic vulnerabilities, and the disruption of the individual's ego functions. Thus far, what predominantly colors the treatment of the issue of paranoiagenesis in organizational life seems to be its linkage with paranoid elements in the personality of the leader, which are perceived as what contributes most to the paranoid culture of the organization (Kets de Vries and Miller 1991, Kernberg 1994). In order to gain a deeper understanding, however, of paranoid regression and paranoiagenesis, and more generally—of psychopathologically informed disturbances in organizational life—we need to go beyond merely personal or individual psychodynamic considerations. The vicissitudes of pathological narcissism, for example, depends upon an interpersonal (e.g., familial) as well as social milieu and institutional climate, which determine whether, and to what extent, it will be contained and modified or turned into malignant narcissism. In much the same way, paranoia entails a social component. The Schreber Case comes to mind in this connection: it illustrates the depth of understanding of intrapsychic homosexual dynamics, ingeniously deciphered and spelled out by Freud (1911). It also contains, however, the interpersonal facet of the cruel, oppressive, and restrictive physical and educational measures imposed by Schreber's father (Niederland 1974). Still beyond this, however, it culls up the particular social climate that gave rise to, enabled, and condoned such unusual and sadistic methods of child rearing. A sadistic cultural climate of this kind is intrinsically bound up with the ways in which authority and power are taken up. In turn, the wielding of authority and power plays a central role in the culture of organizations. We may thus depict the etiology and emergence of paranoia as following a circular course: from individual intrapsychic to multiple interpersonal psychodynamics, to the social and organizational level, and back again to the individual (see Figure 1).

At all three points in this circle issues of authority and power—and hence of their corruption—feature prominently. Corruption has moralistic connotations, and is not often thought of in the context of wielding authority and power. I suggest that social corruption has a significant role in institutional and organizational regression. In fact, I would argue that corruption within the social and organizational realm is the equivalent of psychotic regression in the paranoid individual.

Let me explain and expand this statement. Individual paranoid regression stems from the breakdown of the links between perception and interpretation of social cues under the impact of dangerously intensified internal needs and aggression. This, as I have already suggested, severely undermines the capacity for exercising accurate social judgment. Analogously, corruption in the social realm is linked to the presence of chaos and fragmentation. It particularly involves, on one hand, the breakdown

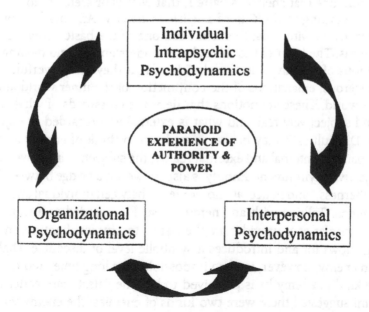

Figure 1: Interrelatedness of Paranoid Experience of Authority and Power at the Individual, Interpersonal, and Organizational Psychodynamic Levels

or outright destruction of the links between manifest social needs and aims, and on the other hand, underlying shared beliefs and expectations (values) concerning the ways and means of their attainment. Corruption thus implies that needs, wishes, and desires may be satisfied in ways not sanctioned or regulated by consensual, observed, normative expectations, codified as moral values and superego guidelines. The result is that social norms—such as "Thou shalt not kill!"—are no longer intrinsically connected to the values of the group. In a corrupt culture, killing and abuse become possible, along with the possibility of being killed by the other, who is thus defined and regarded as an enemy. Social corruption thus stems from and further cultivates the paranoid outlook and posture. Where "anything goes," everything must be feared and hated.

THE DISCOVERY OF ENEMIES

One of the clearest manifestations of paranoia is the discovery of enemies, both real and (more typically) imagined. The term "enemy" is in itself a paranoid definition of an other: "enemy" designates "an unfriendly or hostile person; one that cherishes hatred, that wishes or seeks to do ill to another," according to the *Oxford English Dictionary*. An enemy may also, however, dwell within us; this is indeed one of the basic lessons of psychoanalysis. The heart of the difficulty of understanding and dealing with the notions of "enemy" and "enmity" is in that they are powerful, not to say dangerous, emanations of the conjunction of the inner world and the outside world. These are notions that lie at the crossroads of what is psychic and subjectively real, and what is external and regarded as objective reality. Difficult as it may be, we must learn to think of enmity as a *creation*, spanning internal and external reality, the subjective inner world and the objective environment. Enmity is also, however, a bridge between "self" and "otherness," and hence, at another level, between individual and group phenomena. Talking with an enemy is usually regarded as a significant advance insofar as it provides an alternative to physical fighting; instead, talking allows for and introduces a symbolic level of discourse. Dialogue with an enemy, however, is often impossible for a long time, and depends on the kind of enemy he is perceived to be. The Palestinian leader Feisal Husseini suggested there were two kinds of enemies: the enemy you talk to, and the enemy you don't talk to. The dramatic handshake of Rabin and Arafat, viewed across a shrinking world even by people far removed from the actual conflict with a mixture of hope, disbelief, and astonishment, marked the instantaneous transformation of the enemy one does not talk to into an enemy with whom one does. What distinguishes the enemy we talk to from the one we don't? And how can we turn an enemy we don't discourse with into one that we do?

MEETING THE ENEMY: INTRA- AND INTERGROUP RELATIONSHIPS

In addressing these questions I hypothesize that the intrapsychic enemy becomes a realistic enemy when he or she manifests himself or herself in social reality, and that the main arena in which this takes place is that of intra- and inter-group relationships. My conceptual views in this connection are based largely on Bion's (1961) work in groups, and on numerous observations of these processes within Tavistock-type Group Relations Conferences, which are designed to study group processes in the tradition of Bion and others.

The Concept of Boundaries

One of the notions that evolved out of this work is that of being positioned "on the boundary." The concept of a boundary is in and of itself a creative and evocative one: it plays simultaneously on a number of levels, ranging from the concrete to the abstract, from the real to the symbolic, from the physical to the conceptual. This clarifying and yet evocative aspect has made boundaries such a useful tool in numerous areas.

Boundaries occupy a central position in psychoanalytic ego psychology and in systemic models of group and organizational behavior. They involve notions of strength and permeability, of rigidity and elasticity. Most importantly, there is usually some question about the degree of clarity of their definition. Boundaries, however, are also points of encounter, where different parties can and do meet. Boundaries sometimes allow, or include, a certain amount of "no man's land" not clearly under the jurisdiction of any one party. Often enough, such a no man's land is precisely the territory that encounters and testing-of-limits take place in without the danger and risk of all-out war with full responsibility and consequences. Psychoanalytically speaking, this is also evocative of transitional space and transitional phenomena as described by Winnicott (1971). Elsewhere, however, I have described the different deployment of boundaries in relation to underlying experiential dimensions, depending upon the experience of the self and the object as either "doing" or "being" (Erlich 1990). I pointed out that boundaries between self and object might be drawn sharply, to emphasize their delineation and separateness. They may, however, also encompass the self and the object that are experienced as a merged and fused "oneness" (a state often erroneously conceived of as necessarily signifying the absence of boundaries).

It is actually useful to think of boundaries not as well-defined, razor-thin lines that cannot support or contain life, but as gray areas or untamed territories in which a great deal of action and significant living actually take place. This may often bear the character of some variety of "play," in the sense that it does not lead to immediately and directly binding consequences in well-defined areas of living. Such a boundary, or better yet, a frontier-area, provides a good deal of elasticity and permeability. Moreover, it can give birth to and support what is creative, novel, and psychologically pertinent. But not only positive creative aspects of life have their roots here; negative creations, such as enmity, are also fundamentally linked to the psychological transactions and creations at the boundary. It is this area and the life that takes place in and around it that I have in mind when I speak of the enemy as being created and come to life "on the boundary."

Large-Group Dynamics

If we shift our focus and consider for a moment the dynamics that take place in the large group, we find that enmity occupies a pivotal role in it. One of the centrally important maneuvers in the large group is to mark the enemy and then relate to him as such. Splitting the large group into subgroups and splinter-systems unfailingly does this. Fragmenting of the whole seems so natural, and occurs so frequently and swiftly, that it is often extremely difficult to note and follow. Large-group divisiveness may well be the equivalent of individual intrapsychic splitting of the whole bad object, in order to assimilate and subjugate it. The governing wishful fantasy, however, is to bring about peace and tranquillity—the yearned-for state in which this unremitting, impossible, and difficult frustration will finally stop—through one subgroup gaining control over the entire group. Behind the multiple splits and wars against a shifting variety of enemies is thus actually the wish for final and total submersion in the whole, for a state in which the individual will no longer be a problem because of his separate existence and identity.

Enmity within the large group is thus a tremendously fluctuating, treacherous, and diffuse entity. An enemy identified one moment may be totally disregarded the next. Under these conditions, it is impossible to carry on meaningful discourse with either friend or foe. This constant internal shifting and fluidity makes the large group so dangerous. Its inner instability allows it to be suddenly and irrationally tilted in the direction in which an enemy is identified. The discovery of an external enemy brings about the momentary stabilization of the large group, and hence an alleviation of its tremendous inner tensions. Clearly, this makes the large group extremely vulnerable to paranoid states—to being manipulated into seeking and destroying real or imaginary enemies. It should be noted that, again, the enemy takes shape on the group's boundary, be it its physical, geographic, or ideological boundary. In this boundary-area of the large group, we actually find many different sorts of enemies: barbarian invaders and military threats, religious fanatics and false Messiahs, converts into religion and out of it, and political reformers bent on changing the group. These are indeed often referred to as "fringe groups" and serve as the object of hatred and animosity. The large group's own leaders are, however, also on the boundary, as all leaders always are, and can easily and momentarily be turned into its enemies. To cite a totally different situation: the psychotherapist or psychoanalyst in the transference situation is also on the boundary for the patient, and especially so when delivering interpretations. This may explain some of the enmity directed at individual therapists, as well as group therapists and consultants in both small- and large-group situations.

Otherness and the Stranger-Enemy

You may well ask why this is so. Why is it that whosoever appears on the boundary is potentially an enemy? The answer must have to do with the history of our early experiences of boundary formation, with the most primary and archaic recognition of otherness. The first manifestation of the other, and the anxious and startled reaction to his appearance, is contained in the phenomenon of stranger anxiety, universally encountered at around 8 months of age. The very appearance of the stranger's face—always surprising and unexpected—arouses existential fright and anxiety. This "other-stranger" who provokes this stranger-anxiety is frightening because of his or her otherness. He or she appears at the very moment when fusion with the mother and the experience of blissful merger with her becomes a nearly conscious source of pleasure and security. The appearance of the stranger threatens to undercut and interrupt this merger. It propels the infant into immediate attention, reorganization, mobilization of forces, and readiness to face danger—in brief, an arousal and anxiety response.

The anxiety response to the stranger is undoubtedly universal. Enlarging on this, we may say that the stranger is the prototype of the internal, psychic enemy that has become a social reality. The stranger represents the archaic threat to destroy our peace, to snatch us out of the calmness that comes through Being—the merger with an other in the experience of simply being alive. In both an historical and contemporary context, there is always great readiness to project onto the stranger this role of the enemy, the "destroyer of the peace." But who is this stranger? The stranger I am talking about is not someone distant and unknown. He or she lives within society, yet is not fully a part of it. He or she occupies the same "boundary position" as the leader in the group and the analyst in the psychoanalytic situation. Taking up this boundary position immediately casts one in the role of the other or the Stranger, readily rendering him or her the natural target of projections of hatred and enmity.

In the course of development, stranger-anxiety gradually turns into recognition of the other's separate and independent existence. This recognition is an important basis for the development and maintenance of mature object relations (Sandler 1977). It has, however, an additional facet: the anxiety in face of the stranger-enemy also serves as a primary, almost reflexive reminder of the limitations and liabilities of the self. In this sense, it also provides a necessary condition for realistic self-definition. Paradoxically, then, the anxiety stirred up in relation to the stranger-enemy provides a needed catalyst for the process of self-definition. To paraphrase, if there were no enemy, we would certainly have had to invent one.

DISCOURSING WITH THE ENEMY: TENTATIVE CONCLUSIONS

Is it possible to find ways of talking, communicating, and discoursing with an enemy? I would like to finish with an attempt, which is bound to be a frustratingly partial and insufficient one, to draw some tentative conclusions from what we have surveyed so far.

Freud thought that the individual could be rationally approached and understood. The group, and particularly the large group, however, renders human behavior primitive and irrational, particularly around the eruption of enmity. He saw it as "a mystery why the collective individuals should in fact despise, hate and detest one another—every nation against every other—and even in times of peace" (1915, p. 288). Some advances have taken place in our understanding since Freud's lines were written under the impact of the First World War, including Freud's own extension of his instinct theory to include a primary destructive/aggressive drive (Freud 1920). I suggest that enmity is indeed an inherent part of the individual human psyche; but enmity is also on the boundary between internal and external reality. It takes on its familiar meaning and shape as a social phenomenon when we meet and work with it at the group, system, and organizational levels. These levels, with their irrational, regressive undercurrents, must, therefore, no longer be underestimated in our theoretical, research, and practice efforts.

This brings us back to the dynamics of the large group. From all we know about large-group processes, even under the relatively controlled conditions of a group relations training conference and with the participation of consultants, there can be only one conclusion: large-group regressive processes are highly lawful and regular. This is so even when the participants are persons who have had the benefits of previous experience and impressive educational and cultural achievements. Therefore, if our aim is rational enlightened political activity, large-group settings and events must be avoided and prevented as much as possible. It is certainly true, and has again been recently demonstrated, that large masses of people can be instrumental in changing the political order. But it is equally true that such mass movements and revolutionary upheavals can also go in unpredictable directions; they do not necessarily always lead to freedom and democracy. Large crowds and mobs were involved in so many revolutions—the French, Russian, Nazi Germany, China, the more recent uprisings in Eastern Europe, and so on. This, however, provides no assurance regarding the eventual outcome, nor does it guard against the manipulation of primitive needs and anxieties by sinister powers to their own ends. Wherever possible—and especially where negotiations are to take place—small groups are preferable to large groups. This may also be true for gatherings that are not manifestly (yet are implicitly) political, such as symposia and conventions. Negotiations between enemy parties to a conflict need not only the small-group format, but also the clarity and firmness of boundaries that guard against premature exposure which reverts the process back into the large group.

The small-group format in itself, however, is also not a guarantee for dialogue. Indeed, the need for dialogue is often glibly and unthinkingly advanced. Our professional and personal biases (and I speak as a psychoanalyst) may well be responsible for the introduction of complications, misconceptions, and even fallacies in this regard. We are trained in dialogue, believe in it deeply, and are deeply committed to searching discourse and discussion. We may therefore be prone to minimize or play down, however, the tremendous importance of the psychoanalytic setting, with its unique combination of strict boundaries and open-endedness, in enabling, shaping, and contributing to the creation of dialogue, and then only after much time and tremendous efforts. Having witnessed attempts at dialogue with "labeled enemies" in professional group settings, I have been amazed and perplexed by the degree and speed with which such "dialogic" sessions become confrontational and coercive. Dialogue is based on the ability to recognize the other's essential and rightful difference; this is diametrically opposed to regarding him or her as an enemy. Where dialogue can occur, the enemy is essentially no longer an enemy.

The still-continuing and difficult process of peace talks between Israel and the Arabs provides an example of how actual contact contributes meaningfully to the reduction of strangeness and projections. And yet, it does not miraculously make them vanish. Furthermore, the actual contact is likely to produce and introduce new and unforeseen difficulties. In this sense, direct contact in itself is no guarantee for the disappearance of enmity and the triumph of reason and peace. There are many other factors involved that interact with and activate the psychological ones. Economic issues provide a cogent and almost ubiquitous everyday example, particularly when real problems are synergic with and activate fantasies of shortage and lack of supplies. The entire range of realistic possibilities must be reckoned with, including the emergence of new and insurmountable difficulties that will emerge as the relatedness to yesterday's enemies declines but the relationship with them develops and deepens.

I have described (Erlich 1997), from a combined psychoanalytic and object relations perspective, two categories that enemies may be conceptualized as falling into: the preoedipal enemy as against the oedipal one. The preoedipal enemy is perceived monolithically and in black-and-white terms as totally bad, mad, and destructive. Splitting, projection, and projective identification mark the relatedness with this preoedipal enemy. The relationship with him or her is characterized by total negation, often expressed in the utter lack of readiness to speak with him or her, experienced subjectively as a physically and psychologically insurmountable barrier. The oedipal enemy, on the other hand, is experienced ambivalently, yet as sharing a commonly valued "third," such as a common humanity, culture, heritage, or language. The relationship with this enemy is marked by competition, fear, and envy, but also with admiration and positive relatedness, and discourse with him is felt to be within the realm of psychological and social possibility.

In closing, it seems to me that these views contain some implications for our discourse with enemies. If we regard the enemy and our paranoid perception of him or her to be an inherent aspect of the human condition and the manifestation of our subjective inner world in the social sphere, we have to set our sights concerning our expectations regarding enmity within human history and discourse accordingly. It may be that the most we can ever hope for is to change the enemy from his or her preoedipal position of total badness and destruction to the oedipal level of rivalry and competitiveness coupled with love and affection. We may thus be able to change the enemy from one that we do not talk to into one that we can and do talk with—certainly not a mean achievement.

Distance, boundaries, and separateness certainly make the task of discourse more manageable, though not necessarily more creative. A clearer view of the enemy's "otherness" contributes much to enhancing discourse. The acceptance of the enemy's "otherness" is a concession of his or her humanness, and of the differences, variability, and individuation of persons and groups; it alone can ensure the development of creative conflict management, if not full resolution, instead of fighting and destruction. "Otherness" provides a basis for a fresh view, and thus for new and creative contact, replacing fantasy relatedness, which fosters the wish to destroy and assimilate the enemy. Above all, it contributes to the retrieving of projections. As long as projective identifications are operating, we have a powerful investment in the other's total badness and evil, which allows us, in turn, to be full of goodness and humanity, yet locks us into a position that enables only paranoiagenesis. To recognize aspects of the other in ourselves is as difficult, if not more so, as having a better integrated, depressive-position image of him or her (Klein 1946), yet it is the only way to ensure that he or she will not go on *being evil on our behalf.* Probably one of the most creative acts we may ever be capable of lies in our potential capacity to experience our enemy as a part of ourselves, while also recognizing his or her existence in their own right, as separate and distinct from us.

REFERENCES

Bion, W. R. (1961). *Experiences in Groups*. New York: Basic Books.

Erlich, H. S. (1990). Boundaries, limitations, and the wish for fusion in the treatment of adolescents. In *Psychoanalytic Study of the Child* 45:195–213. New Haven, CT: Yale University Press.

——— (1996). "Ego" and "self" in the group. *Group Analysis* 29:229–243.

——— (1997). On discourse with an enemy. In *The Inner World in the Outer World: Psychoanalytic Perspectives*, ed. E. R. Shapiro, pp. 123–142. New Haven & London: Yale University Press.

——— (2000). Joining, experiencing, and individuating: ego and self in the group. In *Who Am I? The Ego and the Self in Psychoanalysis*. Volume 5, Encyclopaedia of Psychoanalysis, ed. B. Seu, pp. 128–142. London: Rebus Press.

Freud, S. (1911). Psycho-analytic notes on an autobiographical account of a case of paranoia. *Standard Edition* 12:9–82.

——— (1915). Thoughts for the times on war and death. *Standard Edition* 14:273–302.

——— (1920). Beyond the pleasure principle. *Standard Edition* 18:7–64.

——— (1921). Group psychology and the analysis of the ego. *Standard Edition* 18:69–143.

Jaques, E. (1955). Social systems as a defence against persecutory and depressive anxiety. In *New Directions in Psycho-Analysis*, ed. M. Klein, P. Heimann, and R. E. Money-Kyrle, pp. 478–498. London: Maresfield Library.

Kernberg, O. F. (1994). Leadership styles and organizational paranoiagenesis. In *Paranoia: New Psychological Dimensions*, ed. J. M. Oldham and S. Bone, pp. 61–79. New York: International Universities Press.

Kets de Vries, M.F.R., and Miller, D. (1991). Leadership styles and organizational cultures. In *Organizations on the Couch: Clinical Perspectives on Organizational Behavior and Change*. San Francisco and Oxford: Jossey-Bass Publishers, pp. 243–263.

Klein, M. (1946). Notes on some schizoid mechanisms. *International Journal of Psycho-Analysis* 27:99–109.

Miller, E. J. (1989). The "Leicester" model: experiential study of group and organizational processes. *Occasional Paper No. 10*. London: Tavistock Institute of Human Relations.

Niederland, W. G. (1974). *The Schreber Case*. New York: Quadrangle.

Sandler, A.-M. (1977). Beyond eight-months anxiety. *International Journal of Psycho-Analysis* 58:195–207.

Wells, L. (1985). The group-as-a-whole perspective and its theoretical roots. In *Group Relations Reader 2*, ed. D. Colman and M. H. Geller, pp. 109–126. Springfield, VA: A. K. Rice Institute.

Winnicott, D. W. (1971). *Playing and Reality*. London: Tavistock Publications.

6

Dilemmas of Organizational Change:
A Systems Psychodynamic Perspective
JAMES KRANTZ

INTRODUCTION

This chapter focuses on efforts to bring about major changes in the way that organizations function. These efforts, which often involve altering many facets of organization including structures, policies, procedures, technologies, role design, and cultural patterns are increasingly common as organizations adapt to accelerating rates of change in markets, technologies, and competitive pressures. While such changes may be necessitated by turbulent operating environments, they are also profoundly disruptive both to the organizations and to the people functioning within them.

If ineffective, the impact of such change can be disabling, even disastrous, to the ongoing viability of the enterprise and devastating to its members. Indeed, since Miller and Rice (1967) recognized the management of innovation as a crucial part of management, great attention has been directed to, and an entire field has arisen around, the management of change. My intention here is to explore issues of change management from a systems psychodynamic perspective and, in doing so, to consider the reciprocal impact of psychic and systemic factors on the ability of organizations to implement new approaches.

To do this I consider the interplay between the modes of functioning that people adopt to cope with the experience of change and the way that the change efforts are designed and conducted. Psychoanalytic research has illuminated the importance of anxiety and related defenses both to the functioning of individuals and to the functioning of institutions. Similarly, the research tradition emanating from the Tavistock Institute has enabled us to understand the impact of organizational arrangements on anxieties and fantasies of their members and, in turn, on the kinds of defensive maneuvers they employ to cope with them.

The main hypotheses I would like to put forward is that major organizational change efforts pose great psychic challenges to their members and require, in response, distinctive conditions in order to adequately contain the profound anxieties evoked by such upheaval. And, in the absence of these conditions change efforts are likely to fail, in part because members will tend to employ primitive and destructive defenses to protect themselves from the painful anxieties and fears that attend disruption and turmoil.

Periods of change in organizations put great strain on the ability of their members to contain their anxieties. The course of change both evokes and is shaped by heightened anxiety. A secondary concern in this paper is with the production and distribution of emotional toxicity as a byproduct of organizational change. By toxicity I refer to primitive mental contents that, when projected and enacted in organizational settings, lead to destructive consequences. In this connection I wish to draw a parallel between psychic and organizational functioning. In elaborating Klein's (1940, 1946) understanding of infant development, Bion (1962) points out that by containing and modifying the infant's destructive, envious impulses, the mother "detoxifies" them. In a similar vein, my effort here is to describe how failed containment, in organizational terms, also leads to the production and distribution of the destructive impulses I refer to as "toxicity."

The intensity and rigidity of this toxic cycle varies widely. This paper also addresses the relationship between the levels of toxicity produced and the conditions surrounding change efforts. Through an attempt to elaborate these factors, and the different qualities of change efforts, my hope is to add understanding to the management of change and to the factors that mitigate or exacerbate destructive emotional processes evoked in the course of major organizational change.

THE PARADOX OF CHANGE

Even under relatively stable conditions organizations must cope with inherent tendencies toward psychological regression in their members. This is primarily for two reasons that have been explored in detail by researchers of systems psychodynamics. One concerns the anxieties evoked through contact with the tasks themselves. Specific responsibilities carry symbolic meanings that resonate with deeply held experiences and meanings, stimulating unconscious fantasies and intense anxieties that must then be defended against without, hopefully, compromising the ability to function. The other concerns the psychic challenges posed by the need for collaboration with others—peers, superiors, and subordinates—engagements that also symbolize early configurations and relationships and consequently evoke the distress and conflicts associated with early life experiences.

Following the work of Menzies-Lyth (1960) and Jaques (1955)[1], organizations develop modes of operating that, in some measure, function to help people defend themselves against the anxieties and painful feelings that are stimulated in these ways. While these structures, policies, cultural patterns, and other modes of operation—coined "social defenses" by Menzies-Lyth— help members protect themselves against painful feelings and conflicts, they also affect the organization's ability to function. As with psychic defenses, social defenses operate on a continuum between sophisticated, competence-enhancing adaptations and debilitating forms that can impair or even cripple an organization's capacity to function or innovate effectively.

Hopefully, an organization's social defense system will support the capacity of its members to function effectively by helping them contain, and put into useful perspective, the more primitive fears and anxieties evoked through membership and confrontation with complex tasks. Otherwise, people will rely on primitive defenses to protect themselves from the anxieties that arise from splitting and the persecutory atmosphere established when the resulting bad objects created by splitting and projective identification populate the environment.

Reverting to splitting, denial, and projective identification to cope with distressing anxiety leads to genuinely disturbing and psychically threatening organizational environments. Bad internal objects and impulses may be put into particular members or subgroups as a means of getting relief. Where the qualities of thoughtfulness and collaborative competence give way to blame-ridden, rigid, concrete thinking, an escalating downward

---

1. Jaques (1995) has since recanted this viewpoint, arguing that the psychodynamic underpinnings of organizational life do not provide a useful perspective for understanding organizational functioning.

spiral of fragmentation and persecutory functioning can come to dominate—and paralyze—an organization.

Effective change requires sophisticated effort—diagnosis, conceptualization, planning, implementation, and so forth. Yet it is the very features of organizational life that protect them from intrusion of primitive processes—its social defense system—that are at the same time being dismantled. Just as Menzies-Lyth has shown how an important source of resistance to change is the reluctance of members to give up features of organized life that help keep painful anxieties at bay, organizations undergoing major change can lose the capacity to contain primitive emotional states as social-defense systems are dismantled. Consequently, efforts to innovate confront organizations with a paradox of change: change undermines features of organizational life that foster the very qualities of functioning required to make change succeed.

This dilemma takes on an even sharper meaning in light of the amplified psychic challenges posed by change. Adding to the preexisting sources of regressive anxiety and the inherent pulls toward primitive defenses, the anticipation and/or reality of change can be experienced as catastrophic (Bion 1970) because it disrupts established modes of behavior, traditional attitudes, and established relationships. Both loss of the familiar, with its containing functions, and prospects of a more uncertain future, with its new adaptive requirements, elicit profound anxiety.

Periods of change, then, are characterized by heightened anxiety and fear coupled with weakened capacity to contain potentially disruptive emotional states. Among the most distinctive challenges of managing change involves that of creating conditions that help people cope with distressing transitional states that change efforts create and, in particular, doing so in a way that protects the ability of the organization, and its members, to function effectively. Special measures are required to provide appropriate containment during the transition from one approach to another. Since the more overt disarray and disorientation is accompanied by, as it were, an interim phase between the containing capacity of one social defense system and its successor system, organizations are well served to develop approaches to containment that are specific to the transitional period of change.

Eric Miller (1979) addresses this dilemma in relation to the consultant's role by suggesting that consultants make themselves available to contain heightened dependency needs of client systems in the intervening period between the loss of established social defenses and the consolidation of new ones that are reflective of and in alignment with the newly adopted approach to organizing work. A more primitive, and ultimately destructive, manifestation of this can be seen in either excessive dependency on consultants who at times seem to take control of their client organizations, in rote adoption of approaches and ideas articulated by "management gurus," or in rigid adherence to popular ideologies and fads.

What are the qualities of change efforts that can help people function in spite of heightened exposure to anxiety-producing factors? Thoughtfully developed approaches to change often include features that fill the "social defense vacuum" and support members' efforts to protect themselves. Efforts to provide containment fall along a parallel continuum to that of ordinary social defenses: some promote a more integrated, mature, and sophisticated approach to coping with the emotional challenge while others evoke more primitive responses that rely on splitting defenses.

For example, among common elements of change efforts that can take on social defense functions are:

- Transition planning structures that are effectively authorized to manage the complex issues that arise in the course of change.
- Outplacement support helps moderate both the persecutory anxieties stimulated in relation to some people losing their jobs as well as moderating the guilt of those continuing with the organization.
- Articulation of vision, purpose, and goal. This can be done in a way that helps people take an integrated and realistic stance toward change and the future or in a way that is based on glib clichés and grandiose platitudes.
- Communication strategies aimed at providing information.

When primitive modes of functioning go uncontained and unchecked, the environment will be marked by the corresponding kinds of behavior that lead not only to ineffectiveness but also to production of both idealized and despised objects as the consequence of heightened splitting. Projective identification serves as a kind of psychic distribution function employed to establish patterns of intergroup relations that externalize and enact the psychological splitting process as organizational splits. Bad objects that are created, pooled, and then distributed according to covert political principles can be viewed as the toxicity created by the process of organizational change.

A BASIC FRAMEWORK

I would like to put forth the proposition that organizational change efforts can be categorized in terms that are roughly analogous to the states of mental functioning identified by Melanie Klein (1940, 1946) and since elaborated by clinicians and researchers working within the tradition defined by her approach to understanding human functioning. By analogous I mean that various approaches to change foster patterns of defense that parallel those described by her categories and that they exemplify stances toward thought and interaction that correspond to the modes of functioning she delineated and referred to as the depressive and paranoid-schizoid positions.

The depressive mode is a state of mind in which one maintains contact with the full texture of inner and outer reality, where one can mobilize resources to confront these realities effectively, to collaborate in a sophisticated fashion and learn from experience. When operating in this mode, managers bring an integrated frame of mind to complex problems, assess reality from multiple perspectives, understand realistic opportunities, and take actions that are effectively related to reality. It is a state of mind that also enables people to take responsibility for their actions rather than externalize unwanted "parts" or emotional states.

In the depressive position one can think. The impact of the depressive state of mind on managers was elaborated upon in Lapierre's (1989) paper describing how managers exercise power. Those functioning in a depressive mode are more realistic, less grandiose, and able to achieve what he called "relative potence" growing out of a grounded appreciation of both the complex forces that constrain any change effort as well as the real authority that is vested in their roles. He was concerned with issues of psychological development and maturation of individuals in managerial roles.

Both individuals and social institutions must cope with anxieties that arise specifically in relation to the depressive position, chiefly having to do with acknowledging the impact of one's actions on others. Since management practices in general, and change efforts in particular, entail significant aggression, the defenses employed against the remorse and guilt experienced in the depressive position is an important variable in understanding organizational dynamics.

The paranoid-schizoid mode is characterized by efforts to alleviate disturbing anxieties and feelings by relying upon primitive defenses, principally denial, splitting, and projective identification. Bad persecutory objects and unwanted impulses are split off and externalized. Operating from this state of mind leads to highly compromised functioning because it engenders rigid, concrete thinking, blame, idealization, massive projection, persecutory frames of mind, and diminished capacity for reality testing. The managers Lapierre studied that operated from a predominately paranoid-schizoid frame of reference were grandiose in their aims, unrealistic in their expectations, and ultimately ineffectual in their efforts.

Where this analysis diverges from Lapierre's is in considering the impact of, and interaction with, organizational arrangements and systemic variables on the states of mind, or modes of functioning, that predominate in organizational life. Rather than taking the organizational setting as a stage, so to speak, where individuals express their character, my effort is to consider how different approaches to change foster and elicit different modes of psychological functioning. From this perspective I am attempting to identify approaches to organizational change and innovation that correspond with, and mutually reinforce, these states of mind.

In drawing the parallel between types of change efforts and the two modes of psychological functioning, I find it most useful to think of change efforts, as existing along a continuum that describes the extent to which they are thoughtful, sophisticated, and effective. At one end of the continuum are efforts that resemble the paranoid-schizoid mode, both in terms of how they are conducted and in terms of the type of behavior elicited. For purposes of discussion I will label this type of change effort "primitive." At the other end of the continuum are change efforts—called "sophisticated" here—that resemble the depressive position, where higher-level functioning is supported and the anxieties attending deep change are sufficiently contained to prevent the emergence of destructive disarray or scapegoating.

A good example of the difference between "sophisticated" change efforts and "primitive" endeavors can be seen in the orientation toward the future held by members. In the "sophisticated" (corresponding to the depressive position) stance people are able to adopt a hopeful attitude toward the future, tempered by a sober appreciation of the challenges involved in achieving new approaches. The disturbing disarray, confusion, and uncertainty associated with change is manageable in relation to a positive image of the future that makes sense, both in terms of providing continuity with the past and in offering a plausible image of desirable possibilities.

In contrast, an indicator of primitive change efforts is a dual, split image of the future. On one hand there is an idealized, perhaps even utopian, conception of a grand, new approach to organizing work, often described in language that borrows from the current management fad and can even have little to do with the actual work of the enterprise. On the other hand, there is rampant cynicism and despair about the organization's prospects for implementing meaningful change.

The internal anxieties and impulses associated with the bad, devalued future are often split off, pooled, and then projected into certain groups who then become increasingly "resistant" or doubtful about the change endeavor. In these circumstances there is often strong pressure to publicly voice unquestioning support for the idealized image; openly expressed doubt or criticism is considered disloyal. Nonideal aspects of the emerging reality are unnameable, they are frequently denied by leaders, and attempts to address evidence of the nonideal leads to attacks on thinking such that learning and refinement cannot occur. In the most corrosive situations, signs of decay are hidden behind ritualized meetings and empty planning activity.

I believe that it is also possible to further divide primitive change efforts into two subtypes, what might be called "persecutory" efforts and "grandiose" efforts. In persecutory-type efforts managers and leaders of the change feel that the changes are being imposed upon them and that while they are not in accord with the changes they must implement them. The second type of primitive effort, labeled grandiose, are characterized by wildly expansive aims and heroic idealization and self-idealization of the change leadership. The following cases, drawn from consulting projects, are offered to illustrate these two types of primitive change efforts.

The organizational variables that shape and effect this capacity—or alternatively contribute to the persistence of more regressive functioning—are my central concern. Though useful for identifying patterns and shaping consulting approaches, the obvious dangers of oversimplification must be noted. In reality, change efforts fall somewhere on the continuum between the two poles I have identified; they are rarely static, often shifting where they exist along this hypothetical continuum.

For example, while the anxieties that accompany efforts to bring about major change in organizations seem invariably to evoke primitive processes, people's ability to recover their wits and avoid getting rigidly stuck in primitive defensive postures is a sign of sophisticated change efforts, and also a predictor of success. A hallmark of sophisticated change efforts is, in my view, the capacity of people to recognize and respond to those moments in which they have reverted to more primitive—and potentially destructive—modes of operating, and to think about them rather than simply enact the fragmented view of the world that primitive states of mind engender.

The following case vignettes illustrate severely disturbed efforts to bring about major change in organizations. They have been selected to illustrate both sub-types of primitive change efforts—persecutory and grandiose efforts—and offered in part to exemplify how ill conceived and poorly conducted change efforts can readily elicit highly disturbing and disturbed phenomena.

Case A: Persecutory Change at Micro

The company—called Micro here—developed educational software for large publishing companies. It had a strong reputation for its products and ability to deliver large customized programs in a timely and flexible manner using a sophisticated proprietary software engine that it had developed. Financially, the company had done extremely well. Revenues had grown dramatically and the company managed to increase its profits to a point where they were among the best in the industry while maintaining its software platform, which required constant engineering development to keep current the technologically dynamic environment.

As is common with successful start-up ventures, the founders imbued the company with excitement. Micro had developed a pleasurable and attractive culture of work with an informal, intelligent, and lively atmosphere characteristic of so many Silicon Valley companies. The offices were relaxed and comfortable. The young work force was animated, bright, and talented. And there was a great pride of craftsmanship grounded in a strong dedication to the underlying educational mission of the organization. Micro, founded by educators, placed strong emphasis on the sophistication of its educational content.

The change that I want to focus on began when a much larger software development company—called Emblem here—purchased Micro. Micro's management was faced with transforming the organization into a division of the new company. Emblem, though it too developed educational software for children, sold its products into the retail market rather than developing customized programs for large-scale educational publishing companies. As a result its software development practices were fundamentally different from Micro's.

Micro's managers were active participants in the acquisition of their company by Emblem, in some cases signing lucrative employment contracts. The entire team remained in place and accepted responsibility for transforming

Micro into a successful division of Emblem. Though concerned about the implications of joining a larger, and differently focused, enterprise the management team was attracted both by the potential financial rewards and the opportunities afforded by being part of a large company.

The Emblem executives and managers were eager to align Micro's operating methods with Emblem's. In particular, they felt Micro's "elite" orientation was excessive. In contrast to large educational publishing companies, Emblem's retail customers were less interested in the didactic and theoretical integrity of the educational programming, but rather the excitement and interest that the programs generated. Similarly, Emblem regarded Micro's cherished software engine as just another delivery system and had no interest in either owning or maintaining an expensive proprietary platform since they did not see it as a differentiating factor in Emblem's marketplace.

As the work of integration began Micro's executives received messages that conveyed the business perspectives of Emblem, an orientation that posed a challenge to their sense of identity and sources of self-esteem. Not only did Emblem place little value on Micro's educational and engineering sophistication, but what emerged as Emblem executives learned more about Micro's functioning was a belief that these qualities were counterproductive. To the Emblem executives, investment in educational sophistication, and the proprietary platform looked like "fat" in the system—unnecessary expense that added little or nothing to the value of the end product.

Threats to the identities of Micro staff involved in this transformation were apparent on both personal and organizational levels. For example, upper management faced the prospect of changing from being senior executives of an elite entrepreneurial software development firm to being middle managers of a mass-market software company. Or, in another instance, the focus of development efforts was shifted away from educational sophistication, didactic integrity, and consistent application of cutting-edge learning theory to market appeal, speed of development cycles, and efficiency.

Adopting this new approach required changing many facets of the organization. Most dramatic was the expectation by Emblem executives that the Micro programmers and developers would account for their time much more systematically. Formerly developers kept track of their own time and managed it flexibly, now the Emblem executives wanted them to make sure they billed at least 40 hours/week against active contracts Micro had with its clients.

How this element of the change was introduced is revealing. The new system was imposed by Micro's managers who, at the same time, disavowed responsibility for it. In effect they told the staff: "Emblem is making us do this awful thing to you." Unable to bear the guilt and remorse for doing this, the management team engaged in defensive splitting which cast the new Emblem leadership as mean-spirited and uncaring while they themselves were compassionate and loyal, though also hapless victims.

By splitting off their own responsibility, the managers were able to deflect the anger and rage of their staff toward the new Emblem owners, and in doing so maintained an overt sense of solidarity and harmony with them. The rage and sense of devaluation felt by the Micro executives, who denied their responsibility for this transition, was defended against in ways that led to dysfunctional conflict on the client boundary and paralyzing demoralization amongst the programming staff.

I entered the system at the invitation of a member of the Micro executive team to help them think about how they might address what they had come to see as problems with "morale" at Micro. When I met with them, the vice-president for programming responded to my inquiry about the situation they had in mind when they asked me to come by telling of a recent meeting with their most important client that had gone very badly.

Negative feedback from the client had, according to him, been another blow to the project team's morale. Not only was the client dissatisfied with aspects of the work but they were angry and confrontational during the meeting as well. The meeting ended poorly without any resolution or without the kind of dialogue that set the stage for more affirmative problem solving. The vice-president for technology then spoke about the way in which he felt he and his chief engineer had been unable to represent the technology issues during this meeting in an effective or consistent manner. What emerged was that they had brought an unresolved disagreement to the surface during the meeting and each had, in effect, tried to win by getting the client to ally with their position.

Surprisingly, they spoke of this as a simple glitch that had deleterious consequences on the already mounting morale problem rather than seeing the meeting as an expression of meaningful organizational dysfunction. In other words, the unhappiness of their (major) client was viewed by them as another blow to morale rather than as an indication of dysfunction.

To my mind this was an important clue to problematic issues. On further examination, the vice-president of technology expressed some surprise and curiosity at the way that he and his associate had brought this unresolved conflict to the client boundary in such a raw and ultimately destructive fashion. It didn't make sense to him because it seemed so out of character. He saw himself—and others concurred—as having strong management skills in this area that would have previously led to identification of and creative resolution of this conflict at an appropriate time as a matter of course.

As others on the team began to find bits of their own experience in this vignette, a pattern of deadness in the chain of command started to emerge. The capability of Micro to function was being impaired, in part, by deadness in the chain of command that was emanating from the emotional withdrawal of the Micro executives. Furthermore, a pattern of performance problems and dysfunction was framed as "morale problems" in the staff, *as if* the disquiet and withdrawal resided in *them*. Interestingly, the data commonly voiced to support the "low morale" hypotheses was, on examination, rather weak since there had been only a small increase in turnover. Perhaps the exaggerated sense of turnover represented their projected guilt, expressed in persecutory fear of being punished by their staff abandoning them.

This is not to suggest that "morale problems" were nonexistent; many of the staff were unhappy with the direction Micro was taking. Nevertheless, whatever sense of loss and devaluation was felt amongst the staff, it became clear that in addition to projecting the aggression and cruelty onto the Emblem executives, the Micro executives split off and projected into their staff their own split-off sense of powerlessness and despair, amplifying whatever "morale" problems existed.

Since severe splitting and projection leaves one depleted of potentially useful emotional resources, it is not surprising that the Micro executives experienced a kind of emotional deadness amongst themselves and with their staffs. The emotional consequences of incorporating Micro into Emblem had been devastating for the team and they resorted to primitive defenses in order to cope with the rage, sense of devaluation, and guilt evoked by their role in the merger.

Several months later only two of the original management team remained. The others joined Micro's former major client, taking a number of their staff with them, and attempted to create a development environment that reflected the same values and orientation that had been predominant at Micro before it was acquired. The remaining managers and staff, which remained otherwise largely intact, successfully sought out a meaningful role within Emblem.

Case B: Grandiose Change at Eaton

This type of primitive change can be seen in the efforts of the Information Technology (IT) division of a major financial-services firm, called Eaton here. Information management is essential to the success of global trading operations and the Eaton's IT division—considered one of the most successful and enviable among premier Wall Street firms—had an annual budget of nearly one billion dollars, employing nearly two thousand people worldwide.

In spite of its success, the IT division was under increasing pressure to undertake large-scale innovation. Three factors were behind this: Rising tensions and resentments in the firm's business units about the cost, responsiveness, and capabilities of IT was the most immediate source of pressure. Second was realization that shifts in information technology would completely overwhelm the existing application development arrangements in a relatively short time. Maintaining the same level of service to business units with the newly emerging approach to software architecture and engineering would require vastly increased rates of expenditure unless IT adopted radically new approaches to application development that were more consonant with emerging technologies. Finally, the longer-term prospects raised deeper questions about the value of advanced information technology to support trading since the cost of information is in virtual free-fall. In fact, the long-range industry view predicts that trading—the current mainstay—will produce diminishing returns. Even now, many of the major consolidations, mergers, and acquisitions are driven by this logic.

The firm had been frustrated with IT's inability to respond effectively to these concerns. The division leadership had failed to develop compelling ideas or plans to respond to these increasingly severe pressures. The unit was viewed as lethargic, unresponsive, bloated, and driven more by inertia and established procedure than business logic or strategic purpose. The firm's chairman, himself under increasing pressure from the business units to take action, put one of the most successful traders in Eaton's history in charge of IT. Hopefully, by putting IT under someone from the business units, the division would become more aligned with the business units and managed in closer attunement to the needs and wishes of the firm's core businesses.

While the new head of IT had no experience of managing, he had been a brilliant trader of great renown and had been a pioneer in utilizing new technology to attain advantage over the competition. At 34, Ted had accumulated a vast fortune, achieved acclaimed status within this premier world-class firm, and was charged with turning IT into a more efficient and responsive unit, and one poised to confront the emerging challenges facing information technology in the securities industry.

At some level, however, the choice seemed absurd, conveying a simplistic and unrealistic picture of IT and its management challenges. Even keeping a one-billion dollar global information technology organization going, let alone leading it through major innovation, requires sophisticated management skills and an intimate knowledge of its work. Installing an inexperienced manager with no knowledge of application development or of the requirements of providing large-scale information services seemed to express a demeaning devaluation of IT with sadistic overtones. Given the complexity of the situation, and the sophistication required to meet this challenge, sending a young hero with no particular skill or knowledge of the area to "tame the beast" seemed quite omnipotent and grandiose in its conception.

He set out to transform the organization. With a new senior team that he brought together he began to develop ideas about the required directions for change, beginning with new ideas and approaches gleaned from pioneering experiments at leading-edge companies—"best practices" in today's jargon. What emerged was a "vision" constructed of labels, descriptive terms, and catch phrases offered as prescriptive remedies to the problems in IT. In part because the evolving image was not imbued with a deep understanding of their actual work, this image of the future was neither compelling nor plausible to the staff. In fact, the image he articulated turned out to be largely unintelligible or confusing for the vast majority of IT professionals. While Ted's image of the future had no credibility with the Eaton IT professionals, he and his closest associates regarded the staff's lack of comprehension and disengagement as backward thinking, resistance, and sabotage.

As the new team got more excited about its grand future and about the glory that would accrue for transforming IT into an example of a cutting-edge development organization, they found it very difficult to get the others to join their ideas and got intensely frustrated and angry about the seeming impossibility of making meaningful progress. Intensifying the sense of impotence and disaffection was the absence of the traditional mechanisms of direction setting and integration of activity that are typically relied upon to bring about change. The IT organization was composed of highly autonomous, loosely coupled development groups. The usual pathways of hierarchical delegation and accountability were nearly nonexistent since the organization was structured on the partnership model, like the business units. Unlike the business units, however, the work of the IT division required complex collaboration and sophisticated integration of disparate functions.

Nevertheless, the primary organizational status was linked not to functional role but to status and level within the partnership. The incentive system was centered entirely on the partnership status system rather than the task system. What mattered was promotion up the various rungs of the organization to partnership and, eventually, managing director status. The influence exercised by managers had little to do with their functional roles or task authorization but with their partnership status. Complicating matters was the fact that while yearly bonuses were largely set by one's direct report, advancement was determined by each level deciding who, two levels down, deserved promotion.

This created a fiercely political environment which was grossly misaligned with requirements of task and function. In fact, as the sense of estrangement and conflict intensified between Ted's senior group and the majority of the organization, junior staff people who were aligned with Ted and his team started getting punished via the promotional process. The situation became progressively disruptive and dangerous for mid-level professionals. As Ted and Derek lost their credibility, the juniors who had joined them in their efforts to change IT were decisively punished by other partner-level professionals who embodied the status quo and who had never emotionally joined with the vision set forth by Ted and Derek.

More importantly, no structure existed for effecting innovation—without established patterns of delegation and accountability it was virtually impossible to bring about meaningful change. Curiously, there seemed to be very little authority. The organization depended on a system of personal affiliation, informal networks, promotional competition, and tradition to steer itself. Introducing an accountability system would have entailed a profound, and acrimonious, change in its own right. And since Ted's (and his team of managing directors) esteem was so closely tied to the partnership framework, at every choice point he turned to exhortation and inspiration as the mechanism of change.

The chief vehicle for this was to be the Quality Group (QG), a small unit that had previously been a collection of functions that addressed issues of quality management, such as software testing and teaching project-management skills to the developer groups. It was a small, low-status group that, under the new leadership of Derek, a managing director brought in by Ted for this purpose, would lead all of IT into a new territory. Derek, like Ted, was a successful trader with no experience in management or IT per se, but someone of great energy, curiosity, and ambition.

The "theory" of organizational change that Ted and Derek adopted was one largely of change by inspiration: the QG would first transform itself into the embodiment of a cutting-edge organization and that the clarity, power, and creativity emanating from the group would be compelling for the larger organization. Derek, an inquisitive individual with eclectic interests by nature, actively continued to search out all of the best practices and cutting-edge approaches to organization, elaborating and amplifying the "vision" of the future with more labels, prescriptive concepts, and highly idiosyncratic terminology. Soon the newest approaches and frameworks became the currency of discussion, planning, and expectation: learning systems, knowledge management, the capability maturity model of software development, causal loop diagrams, lean production, and so forth. Derek's tendency was to come on Monday, excited about the management book he had read over the weekend and how it illuminated aspects of this "model" they were developing became a standing joke. Clichés, gimmicks, and canned techniques dominated conversation and thinking, faddish concepts were commonly relied upon as it increasingly began to seem as if these efforts were somehow "out of touch" with the reality of the QG's work as well as that of IT.

Concurrently, the QG was falling apart. First it fragmented into two groups—those aligned with Ted and Derek, and those who felt they were out of touch, with the latter group growing over time to the point that Derek became largely isolated and ineffectual within the QG.

Derek hired us to help the QG perform more effectively, and to help him realize his dreams for the QG and, ultimately, for all of IT. As we began to work with the group we encountered deeply troubling signs. Throughout, we were struck by the dramatic discrepancy between two unfolding stories that were continually presented: One was an elaborate and grand conception of the QG and its transformative potential. The other is a story of ongoing disarray, conflict, resentment, and ineffectual performance within the QG, and continuation of the standing practices and processes in the rest of IT.

A number of interesting features of our consultation also illuminate aspects of primitive change efforts. We too were "caught" in split projections. Initially Derek and his supporters in the QG were great advocates of our work while his detractors were intensely hostile toward our work. But more telling was the experience of having our work both idealized and devalued simultaneously by Derek and his closest associates.

I often found myself speaking and writing with an unusually articulate clarity and incisiveness. My experience of great perceptiveness and lucidity was so vivid and pronounced that it was clearly, to some degree, the result of unconscious group dynamic processes. In a similar vein we developed an extensive and detailed description of the workflow and decision-making processes of the component elements of the QG at Derek's. Though we expected great reluctance on the part of key people to participate, the opposite was true and we were able to produce an enormously rich and useful bit of learning about the QG and about important aspects of IT. Yet somehow it never seemed to have an impact beyond evoking appreciative and admiring comments.

However, in contrast to the apparent clarity and magnificence of our reports and notes, there was equally strong evidence pointing to deep devaluation of our work. Not only was the thinking we offered "lovingly ignored," but it never seemed to go anywhere. Additional problems arose with billing and payment.

My partner, Marc Maltz, and I developed intense countertransferential reactions to the client system. Often we found ourselves coping with pronounced feelings of irritation, despair, and paralysis. Our emotional responses to the situation alternated between polar extremes. At times we planned to resign from the consultation out of a sense of devaluation; other times we were made to feel that it was our work that would produce the critical breakthrough in the situation and that because of our intervention the grand transformations dreamed of would be possible.

Inspiration and exhortation degenerated into bullying and intimidation as the situation deteriorated. A clear indication that the paronoid-schizoid phenomena were coming to dominate was Ted's posture of harsh, dictatorial control. The more he was confronted with evidence of his inability to control the IT organization, the more he bullied and behaved abusively. He became increasingly enraged, exhorting people to change and resorting to threats and tantrums.

For example, one element of their attempt to create a cohesive understanding and planning framework for application development was the institution of a simple reporting form to track the development process. Many professionals simply ignored them and at one point Ted became so enraged by his lack of authority to bring about even this small behavioral change that he lashed out at a mid-level staff member who hadn't conformed to the new procedure by reducing his bonus compensation by several thousand dollars. On one level this painfully illustrated the extent to which Ted had to go to bring about even minute change, given the existing organizational arrangements.

But in terms of the focus of this paper, it also highlights how the efforts to create a new world of empowerment, creativity, and collaborative innovation in fact produced a highly punitive environment of suspicion, persecutory anxiety, and contemptuous sabotage.

Another cardinal feature of grandiose change efforts that I feel is well illustrated in this case is the devaluation and demeaning of the past and current work of the organization. An entirely new departure is called for, one that is free of the imperfections and shortcomings of the past as they are manifested in the current arrangements. As with all idealization and self-idealization, the split-off denigrated parts get located and enacted elsewhere.

FEATURES OF SOPHISTICATED AND PRIMITIVE CHANGE EFFORTS

Moving from a more intensive focus of case study to a standpoint that surveys the broader range of change efforts, the following lists identify qualities that characterize the two basic types of change efforts I am trying to explore. This is not intended to be a comprehensive description of the qualities of these two types but rather a summary of the patterns that have emerged from my own experience.

Sophisticated Change and Depressivity: Effecting significant change in organizations is, to be sure, a daunting challenge and one that requires skill and subtlety on a number of dimensions simultaneously. Organizational change efforts falling into the sophisticated category are characterized by features that are consistent with realistic, grounded, thoughtful functioning and include:

- Genuine investment in structures designed to "contain" and address issues pertaining to the change effort.
- Realistic assessment of the time required to effect significant change.

- Appreciation of how much time people must devote to bringing change about and the impact of this redirection of energy on productivity.
- Respectful recognition of the anxiety evoked by major change efforts and, in particular, recognition that certain segments of people may be significantly disadvantaged and hurt.
- Opportunities for people to acknowledge their complex feelings about such change efforts, including both the depressive and angry dimensions of losing the familiar.
- Toleration of learning from inevitable mistakes and a corresponding ability to make mid-course adjustments as a result.
- Articulation of a plausible and compelling picture of the future that is commonly shared and understood.
- Clarity about how the change effort represents continuity as well as discontinuity—how it is linked to the past.
- Carefully planned and thoughtfully executed, with an appreciation of the human as well as economic and technical factors that intermingle to produce successful outcomes.

Primitive Change and Paranoid-Schizoid Functioning: Klein's paranoid-schizoid mode, characterized by grandiosity, persecution, and inflexible thinking, is a state of mind tending toward obsessional ritual, omnipotent fantasies of control, and paranoid blaming. Managerial actions that are persecutory and/or disconnected from realistic possibility are the hallmark of this state of mind, a state that strives for emotional equilibrium by utilizing primitive defensive maneuvers—chiefly splitting and various forms of projection by which painful, threatening, or frightening aspects of experience are expelled.

Organizational life that is shaped by this mode of functioning is ineffective, dysfunctional, and dangerous. Evacuation of threatening or dangerous elements of mental life produces a toxicity that gets distributed in various ways and can easily lead to severely impaired thinking, inability to learn, paralysis, or destructive scapegoating. Since organizational change efforts are one source of great uncertainty and anxiety they are prone to evoke this kind of functioning or, more precisely, to create pressures for organizations to move toward the paranoid-schizoid end of the spectrum.

The characteristics of primitive type change efforts, in my experience, include:

- Extreme expectations of change in unrealistically short time frames.
- Inconsistent leadership, often marked by changing, and often implausible, images of a sought-after future.
- Clever epithets and superficial ideologies are used to avoid struggle or sidestep the painful difficulties involved in change.
- Grandiose, self-idealizing leadership that is susceptible to the magic elixirs and simple solutions that emerge from efforts to popularize and promote various change technologies. The great complexity of organizational reality and of change efforts get reduced to superficial nostrums or panaceas which, when recited like mantras, stop thought.
- Denial of the human consequences and impact of the change.
- Either the absence of structure to "contain" the change process or artificial structures that are not vested with genuine authority or meaning and are given "lip service" by people who conduct the change process in alternative, often largely covert, fashion.

TOXICITY AND THE DISTRIBUTION OF AFFECT IN ORGANIZATIONS

Finally, I would like to consider how toxicity produced in the course of primitive change efforts gets distributed. Just as Bion (1961) admonished us against forgetting that "man is a political animal" at our own peril, we must be equally mindful that the psycho-social process of group life entails unconscious negotiation over whose needs—emotional and otherwise—will be met. The distribution of affect in group life has been a central topic of concern by students of system-level psychodynamics, a study made possible by the discovery of projective identification. Projective identification is a defensive maneuver by which internalized bad objects and impulses are externalized and then, effectively, absorbed by others. Menzies-Lyth and Jaques pioneered our understanding of the underlying strata of emotional relatedness that stems from the defensive expulsion and pooling of primitive emotional contents in social organizations. Organizations then develop structures and patterns of interaction that support these means of defending against disturbing emotional experiences.

As has been demonstrated by so many researchers, the emotional dynamics of groups and organizations involve tacitly agreed-upon pathways and patterns for the management of complex and challenging unconscious experiences. To name a few that have been explored: how unions can carry the "fight dynamics" on behalf of their larger systems (Rice 1951); student nurses containing the confusion and incompetence that was structured into ward decision-making processes (Menzies-Lyth 1961); or how the projection of incompetence across interdependent units in the U.K. construction industry brings relief to the underlying emotional threats and anxieties although at the cost of disabling the collaboration between the units required to bring about, at least consciously, desired change (Holti and Standing 1997). Another fascinating illustration of this process was developed by Berry (1979) who demonstrated mathematically how adopting specific target-based bonus systems (the most typical kind) effectively shifts risk hierarchically downward.

The case of Micro above illustrates how the executives' intense reactions to the change of ownership structure led to a disabling process of splitting and projection whereby the subordinate staff "contained" the intense rage and discontent. Similarly, with the story of Eaton, the unacknowledged devaluation and denigration that is produced alongside self-idealization and grandiosity found its way into the IT division and contributed to the ultimate failure of many who joined in the effort to bring about change.

Since power entails, to a degree, the ability to define reality, the direction of unwanted emotional elements seems to be usually downward in terms of hierarchy and status. But not solely, especially if we consider the extent to which dependent longings in groups and organizations often lead to situations in which leaders are often reviled, despised, and held responsible for events and outcomes that cannot really be laid at their feet.

One of the defining differences between change projects that tend toward the depressive end of the spectrum and those that tend toward the paranoid-schizoid end is the intensity and rigidity with which one group or another comes to carry the emotional burden of representing the unwanted, dispossessed bits that are evoked in the course of major change efforts. That is, when depressive states of mind and flexibility dominate, untoward projections that come to reside in one group or another do not calcify and become fixed, and thus, change efforts are more likely to be robust.

And in contrast, where a certain group gets laden over time with undesirable projections—scapegoating in other terms—the effort, in my experience, is compromised on several dimensions including the corrosive effects of unspoken guilt and damage to formerly valued relationships; loss of important aspects of the experience of change and adaptation by those who have mentally extruded these experiences, so that the reality and importance of them cannot be accounted for by ongoing efforts to learn and refine the change efforts; and increased cynicism toward authority as a defense against bearing responsibility for the damage caused by projective processes.

One interesting example of this concerns the role middle management seems to be playing in many attempts at restructuring that are built around team-based strategies of organizational architecture. In many instances, I have noted that the course of events leaves the top tier and the lower tiers in far more intact, stable, and well-defined teams than is the experience of the middle. Often times middle management, by contrast, seems to exist in a much more amorphous, chaotic state of disorientation, loss of identity, or loss of clear purpose that resembles the large-group experience in group relations conferences (Turquet 1975).

Is there an unconscious group dynamic whereby middle managers in the midst of these large-scale change efforts become the receptacles for the most unbearable disarray, chaos, uncertainty, and doubt? An hypothesis I developed elsewhere suggests that the emphasis on technology as a more effective way of handling information flows, combined with reliance on distributed and flexible decision making and what is often a contemptuous attitude toward "bureaucracy" can lead to situations where middle management comes to symbolize the features of the "past" that represent the enemy of change (Krantz 1998). When this situation persists it can lead to significant damage both to the organization's capacity to re-form itself effectively as well as to the individuals who must carry these projections.

CONCLUSION

An analysis of this sort inevitably leads to frustrating dilemmas: How can organizations interrupt destructive or debilitating cycles of primitive organizational change? What can be done to help organizations adopt effective approaches to change? And, perhaps more urgently, what can be done to intervene in failing or highly toxic change efforts? As the case illustrations foreshadow, I have no satisfying answer to this question beyond a reliance on the modest power of reflection, what Eric Miller refers to as "holding up a mirror to the client system" (private communication, 1998).

There is no doubt that productive membership in contemporary organizations is calling upon ever higher levels of functioning and greater interpersonal sophistication. At times, the visions of high-performance, team-based settings, communities of practice, or learning organizations are built around utopian and unrealistic images of humanity. Take, for example, the comments of Charles Handy (1996), a leading voice in defining emerging organizational arrangements, about the qualities of a "learning organization: "The learning organization is built upon an assumption of competence that is supported by four other qualities or characteristics: curiosity, forgiveness, trust and togetherness."

The new, utopian conceptions of organizational life that are in vogue now are often bereft of ideas about containment of the primitive, destructive features of human functioning, features that are inherent in organizational life, and possibly exacerbated by the increasing rates of change and fluctuation. Yet in wishing away the destructive impulses and debilitating conflicts that are elicited by membership in work organizations, important generative forces also get overlooked, since the unconscious is the source of creativity as well as of destructiveness. Finding ways to tolerate the discomforting and destructive elements of our experience and ways of linking the raw unconscious forces to our conscious aims seem, increasingly, to be a formidable challenge in creating generative organizational environments.

This then leads to my final point, namely that the analysis forming the basis of this paper might be subjected to a similar criticism. Asserting the importance of building organizational change efforts that support higher, sophisticated functioning perpetuates the tendency to split off, and devalue the primitive strata of experience and, more importantly, to turn attention away from Klein's developmental dynamic. Her framework crystallizes development as an ongoing oscillation between the two phases of mental functioning, with each modifying the other and providing ongoing opportunities to integrate both conscious and unconscious, rational and irrational elements at ever higher levels of sophistication and maturity.[2] The exploration of organizational change that holds this developmental tension at the center of inquiry, as a creative force in itself, remains to be done.

---

2. I am indebted to David Armstrong for this insight.

BIBLIOGRAPHY

Berry, A. J. (1979). Policy, accounting and the problem of order. *Personnel Review* 8:36–41.

Bion, W. R. (1961). *Experiences in Groups.* London: Tavistock Publications.

———— (1962). *Learning from Experience.* London: Heinemann.

———— (1970). *Attention and Interpretation.* London: Tavistock Books.

Handy, C. (1996). *Gods of Management : The Changing Work of Organizations.* London: Oxford University Press.

Holti, R., and Standing, H. (1997). *The complexity of organizational life—how does psychoanalytical thinking broaden our understanding?* Presented at the International Society for the Psychoanalytic Study of Organizations symposium.

Jaques, E. (1995). Why the psychoanalytical approach to understanding organizations is dysfunctional. *Human Relations* 48:343–365.

Klein, M. (1940). Mourning and its relation to manic-depressive states. In *The Writings of Melanie Klein Vol. 1: Love, Guilt and Reparation.* London: Hogarth Press.

———— (1946). Notes on some schizoid mechanisms. In *The Writings of Melanie Klein Vol. 3: Envy and Gratitude and Other Works*, pp. 1–24. London: Hogarth Press.

Krantz, J. (1998). Anxiety and the new order. In *The Psychodynamics of Leadership*, ed. E. Klein, F. Gabelnick, and P. Herr. Madison, CT: Psychosocial Press.

Lapierre, L. (1989). Mourning, potency, and power in management. *Human Resource Management* 28:177–189.

Menzies-Lyth, I.E.P. (1960). The functioning of social systems as a defence against anxiety. *Tavistock Pamphlet No. 3.* London: Tavistock Publications.

Miller, E. J. (1979). Autonomy, dependency and organizational change. In *Innovation in Patient Care: An Action Research Study of Change in a Psychiatric Hospital*, ed. D. Towell and C. Harries. London: Croon Helm.

———— (1998). Private communication.

Miller, E. J., and Rice, A. K.(1967). *Systems of Organization.* London: Tavistock Publications.

Rice, A. K. (1951). The use of unrecognized cultural mechanisms in an expanding machine shop. *Human Relations* 4:143–160.

Turquet, P. M. (1975). Threats to identity in the large group. In *The Large Group*, ed. L. Kreeger. London: Constable.

7

A Large-System Intervention:
The Influence of Organisational Culture

LIONEL F. STAPLEY

INTRODUCTION

The aims of this chapter are to examine the influence of organisational culture; to identify some of the problems that it creates for those working in organisations; and to demonstrate a practical means of analysing this elusive phenomenon. The approach taken will be to outline the theory and then apply it to a large-system intervention in a health-care organisation as a means of showing how that knowledge can be helpful—even essential—in providing the desired understanding necessary to guide our consultation efforts.

Whenever we think of large-system interventions it inevitably means not only looking at the way that individuals and groups relate to each other, but more especially the way they relate to the organisation itself. It also requires an understanding of the way that processes in society affect the dynamics of the organisation. Part of our analysis must, therefore, be concerned with the prickly and troublesome problems associated with organisational culture.

Organisational culture has a significant influence on the dynamics of an organisation, and it is well-recognised and accepted that any change within an organisation will be affected by the organisational culture. Ignorance of the organisational culture or, worse still, attempts to deliberately work against it will render change exceedingly difficult and perhaps impossible. And, even where we are not seeking to influence the organisational culture per se, it will still be important that we understand and work with the culture if we are to have the best possible chance of achieving desired change. This is more especially the case when it comes to large-system interventions where the organisational culture will have a considerable impact on the dynamics under review.

ORGANISATIONAL CULTURE

There is a wide divergence of views and a lack of consensus of meaning about organisational culture that leads to confusion and a lack of understanding. Nevertheless, if we are to work in organisations and particularly if we are attempting to bring about change in organisations, it is important that we have a means of studying and understanding this phenomenon. For current purposes, the theory of culture first developed in *The Personality of the Organisation: A Psychodynamic Explanation of Culture and Change* (Stapley 1996) will be used to conceptualise the intervention process. The main points of the theory are briefly outlined below.

It is postulated that the key to understanding organisational culture lies in an understanding of how culture develops. Knowing how it develops will permit us to unpack it and therefore know how to influence it. In simple terms, we can say that culture develops out of the interrelatedness of the members of an organisation, and what I shall term the organisational "holding environment".[1] The organisational holding environment consists of a physical and social part, which is particularly influenced by factors

---

1. The concept of a "holding environment" was developed by Winnicott (1971) and was considered by him to be vital to the development of the infant. From the beginning of life, reliable holding has to be a feature of the environment if the child is to survive. It starts with and is a continuation of the psychological provision that characterises the prenatal state:

> He was of the view that the mother provides the very context in which development takes place, and from the point of view of the newborn she is a part of the self. She provides a true psycho-social context: she is both "psycho" and "social", depending on whose perspective we take, and the transformation by which she becomes for the infant gradually less "psycho" and more "social" describes the very evolution of meaning itself. [Stapley 1996, p. 29]

> In much the same way that we interrelate with the maternal holding environment, so we interrelate with the organisational holding environment. We use it to supply the same needs as the maternal holding environment and we apply the same affect to it, and create the same defences when it is seen as "not good enough".

such as the leader or ruling coalition, policies, structure, and strategies. In addition, it also includes a psychological part, which is largely unconscious and forms the basic social character of individuals and groups within the organisation, and is based on their past experience.

Being a construct, we need to consider the way the holding environment is viewed and developed by the members of an organisation. I have previously noted that:

> It is by the process of perception that we impose some structure on new input, compare it with a pool of old information, and then either add to it or eliminate it. We can only make judgements about whether we like or dislike something if it is something we know. Our sensations must be completed by some form of appraisal before we can decide whether it is good or bad for us. What we are concerned with is not so much "reality" (whatever that may be), but rather how the human individual develops his or her reality concepts and uses them to live in the world. [Stapley 1996, p. 4]

We experience the organisational holding environment through our perceptive processes and filter them down and match them against the pool of internalised information consisting of our past experience and then develop a construct of the organisation that we may refer to as the "organisation in the mind" (Turquet 1974). It is this construct of an organisation in the mind that members of the organisation interrelate with. It is an object that they create from the perceived view of the organisational holding environment and this leads us to the way that culture develops. Having developed the construct of an organisation in the mind, the members of that organisation then adopt forms of behaviour that they feel are appropriate to them under the circumstances that they perceive are imposed on them by their holding environment.

The construct that we refer to as the "organisation in the mind" is important, but for a deeper understanding of organisational culture we need to go beyond this to an understanding of the organisational holding environment. It is not only helpful but necessary, for current purposes, to view

---

The reader will appreciate that there is no "mother" in the organisation setting. So why is it that members of an organisation perceive them as being in existence and as real as our mother? We identify with the organisation "as if" it were real: it is what might be referred to as an "organisation held in the mind". It is a construct that we identify and treat "as if" it were real. [Ibid. p. 37]

the holding environment as consisting of two parts. In this respect the "ice-berg" analogy previously used by other writers may be a useful way of view-ing things. The sociological part of the holding environment is that part which is exposed or is conscious: I shall refer to this part as the "external holding environment". The psychological part of the holding environment is internalised and largely unconscious: I shall refer to this part as the "in-ternal holding environment". Both parts are more fully explained below.

Starting with the external holding environment: this is basically the sociological holding environment. It includes the formal structures and strategies, the ruling coalition or leader, the organisational tasks—that is, the reason why the organisation exists, the roles of the various members, all forms of knowledge and skills, and the values and attitudes shared by the members. Trist (1990) points out that for each of the members of the organisation these external social objects are regarded by the individual members as their psychological possessions rather than as aspects of them-selves. That is, they exist within them as material that they can use, of which they are partially aware, and that they are able to make available to them-selves by the normal process of recall. In addition, an organisation does not exist in isolation; it is an open system (see Miller and Rice 1967) that interacts with the external world and, depending on the degree of open-ness, this provides some of the external social objects.

Turning now to the internalised holding environment: this is basi-cally the psychological holding environment. Here we are referring to inter-nal objects which are regarded as part of the self and compose the basic social character of the individual. They relate to the deeper character level and derive from the phantasy activity of unconscious systems of internal object-relations. According to Trist, "They act as an internal source of in-fluence on the patterns at a more conscious level and reach into society through them. They may also be directly, though still unconsciously, pro-jected onto various types of external social objects which themselves are then partly fashioned by these investments" (1990, p. 542).

It will be appreciated that while the division into an internal and exter-nal holding environment is valuable for the purposes of understanding, in formulating their view of the organisation in the mind the members of an organisation utilise no such division. However, the important point is that the process of developing this construct occurs both on a conscious and unconscious level, and has rational and emotional aspects. This will become clearer in the following material as the theory is first applied to group rela-tions training conferences and then to a large-system intervention.

Organisational culture is an important aspect of group relations training and provides an experiential learning opportunity for members to gain a deeper understanding of this phenomenon. Rice (1965) was of the view that, "The culture of the conference is its customary and traditional ways of thinking and doing things, which, eventually, is shared to a greater or lesser degree by staff and members alike". He also hinted at the internal and external holding environments when he continued, "It covers a wide range of behaviour—methods of work, skills and knowledge, attitudes towards authority and discipline, and the less conscious conventions and taboos" (p. 43).

He also highlighted the need for managerial awareness of culture when he stated, "At the beginning of the conference, therefore, it is conference management, conference setting, and staff behaviour that have to provide the means by which the basis of conference culture is established" (p. 43). In other words, conference staff and management have the responsibility of providing a "good enough holding environment", one which will enable the members to work at the conference task. Because of the nature of group relations training conferences the holding environment must be one in which aggressive behaviour, expressions of hostility between individuals and groups, can be studied and their effect on decision-making examined and learned about, without their becoming destructive—either of the individual or of the conference. Furthermore, since the task of the conference is to provide opportunities for learning about leadership, group, and organisational processes, the pattern of authority and responsibility in the conference has to be sufficiently explicit to be capable of examination, and sufficiently stable to be able to tolerate critical and even hostile scrutiny.

Most importantly, Rice introduced the notion that, "In any institution, 'cultural congruence', the extent to which the culture 'fits' the task of the institution, is as important for effective task performance as structural fit". Where there is such congruence and conference members experience the institution holding environment as "good enough" the culture will be task-supportive and we shall have growth or progression. Growth always involves a process of differentiation, of emergence from what Kegan (1982) refers to as "embeddedness". Thus, progression means not just differentiation but differentiation and reintegration, a process which is essentially that of adaptation. Subject-object relations emerge out of a lifelong process of development: a succession of qualitative differentiations of the self from the world; successive triumphs of "relationship to" rather than "embeddedness in". However, should the institution holding environment not be perceived as "good enough" by the membership it may result in regression or an anti-task culture.

If the organisation in the mind is viewed in a negative manner and there is not what I shall refer to as a "basic trust"[2] this will almost certainly result in some behaviours that will be influenced by unhelpful psychological processes that have their roots in infancy and which are more likely to stimulate immature behaviour. Thus, if the members of an organisation perceive the organisation in the mind as one which is highly influenced by aggression they are likely to adopt forms of either aggressive or passive behaviour that they feel are appropriate in the circumstances.

THE INTERVENTION

In this part of the chapter I will illustrate and expand on the ideas set out above by applying them to a large-system intervention. The subject matter of the intervention is not really important to this discussion, but by way of setting the scene, I will provide a brief introduction. The purpose of the intervention concerned organisational stress, and the consultancy was carried out by a team of OPUS[3] consultants over a period of about 18 months. What follows is not a full case study of the intervention, but rather a selective and focused use of the material that will help the reader

---

2. "Basic trust" is a further concept developed by Winnicott (1971):

> For the infant to develop there is a need for a "basic trust" in the maternal holding environment and for what Winnicott (1971) has termed "a good enough holding environment". This "basic trust" is developed as a result of the infant's perceived experience of his holding environment. From holding in the mother's womb this extends to holding in the mother's arms. However, what we are referring to is much more than just physical holding. It is about the mother providing boundaries which help the infant to make sense of his world. [Stapley 1996, p. 31]

In the organisational setting, "basic trust" is developed out of the experience of the organisation holding environment by the members of the organisation (p. 38).

3. OPUS (an Organisation for Promoting Understanding of Society) was founded in 1975 and is a registered charity and company limited by guarantee. Its name reflects its aim, which is to encourage the study of conscious and unconscious processes in society and institutions within it. Its members are professionals from a range of disciplines. OPUS undertakes research, organises conferences, promotes study groups called "Listening Posts", and publishes bulletins and papers. It also provides consultancy for client organisations, predominantly in the public and voluntary sectors.

to understand aspects of the intervention relating to organisational culture. Needless to say, other issues were also important, but in order not to detract from the aims of the chapter, these other matters will not be referred to here.

The consultancy was to a health-care organisation that had been in existence for about 5 years at the time of the intervention. It had been developed from an amalgamation of three other units, two concerned with community care and one concerned with mental-health care. It is a large organisation with about 3,500 personnel spread over 50 different sites, covering a wide geographic area. The chief executive, senior managers, and others had worked extremely hard and developed a number of initiatives to try to bring about desired change. However, it appeared that nothing they attempted was having an effect on the organisation. Faced with this highly frustrating experience, the chief executive and senior managers decided that they needed the help of external consultants.

The presenting problem was one of an organisation that seemed to be stuck in a situation which was evidenced by low morale encompassing the following issues: concern regarding job security; increasing work pressure; high sickness levels; passivity; tribalism; and opposition to change. The senior management group had now reached a point where, having attempted their known and trusted ways of managing without success, were now totally frustrated. This point was most adequately expressed by the chief executive who emphatically stated "I just can't unlock it!"

Having obtained the authority of the chief executive and senior managers for the work to proceed a small project-management group was set up for the purpose of overseeing, administering, and liaising with the consultants. This was chaired by a member of the senior management team and consisted of mainly middle managers and staff representatives. As the description of the intervention unfolds, it will be seen that the projections of members of the organisation were such that this group were at times related to as if they were the organisation in the mind, while at other times they were representing the members of the organisation and were themselves relating to the organisation in the mind.

Some early data about the organisation was provided by the project-management group in response to being asked to say how they would like to see things in 18 months time. Some of the responses were as follows: employees being more valued; more trust; to have a sense of working together; to be able to feel more secure about role and organisation; a lack of hidden agendas; to feel less persecuted; to be more in control of things;

less sickness absence; people feeling less tired and responsive to work; to not feel the need to get behind protective barriers; and to be able to seek staff support without feeling that it was a sign of weakness.

It may be helpful at this stage to take a short detour to discuss methodology. *In The Leicester Model* (1989), Eric Miller describes how:

> A still more specific derivation from the analyst's role has been the stress laid on examining and using the transference and counter-transference within the professional relationship. That is to say, the way the consultant is used and experienced, and also the feelings evoked in him or her, may offer evidence of underlying and unstated issues and feelings in the client system: that which is repressed by the client may be expressed by the consultant. [p. 11]

In small group situations this may be exceedingly helpful as one of the means of understanding more deeply the dynamics of organisational behaviour. In large-system interventions where we are working with small groups, such as the senior-management team and in this case the project management team, an analysis of transference and countertransference may be equally helpful.

The way the group relates to the consultant can provide data with regard to how they relate to their holding environment; that is, it can provide data with regard to the culture. However, it may equally provide information about subcultures as opposed to the overall culture of the organisation. In addition, as Eric Miller also notes:

> Jaques drawing on Klein and building on Bion, offered the equivalently useful theory of larger social systems as providing defences against persecutory and depressive anxiety (Jaques 1955); Menzies (1960) was then able to identify the more specific anxieties and associated defences in nursing. [p. 21]

Consequently, in addition to the helpful information that we may obtain from transference and countertransference, it is suggested that we need some additional means of getting access to the total organisation dynamic in order to make the above information available.

Although not in any way overtly stated, a central aim of the consultation concerned gaining an understanding of the perception that the members of the organisation had of their organisation holding environment. The methodology adopted was a derivation of a system called "Listening Posts" that has been developed by OPUS and which have been used over a number of years for gaining an understanding of societal dynamics. In simple terms, a Listening Post consists of a group of at least 12—but sometimes over 40 members—who all attend voluntarily. The task for the members of the Listening Post is to relate to their own experience and pre-occupations of society and then to try to make sense of that experience by formulating some form of hypotheses or statements about that experience.

This approach is further explained by Dartington (1997) who describes how:

> The concept of a Listening Post is based on certain propositions: that people may with some encouragement take up a (reflective) citizen role, mobilising their observing ego; that the dynamics of the group may be such that even a small group may nevertheless act as if it is a microcosm of the large group that is society, so that the themes that emerge through associative dialogue may legitimately be analysed for their societal content.

This idea is further expanded upon by Olya Khaleelee and Eric Miller (1985) in "Beyond the Small Group: Society As An Intelligible Field of Study".

Applying the Listening Post method to this organisation, we referred to it as a "Listening Group", which we felt was more appropriate and more easily understood. This particular event, which was held over two days, was focused on the theme of identifying organisational causes of stress. In all other respects the methodology was the same as the Listening Posts. There was a membership of 36 people who were all volunteers and were a representative slice across the organisation taking cognisance of such factors as hierarchy, geographical spread, and functions. As with any sampling it is never possible to cover all and every part of the organisation, but by careful planning whereby some of the invited members represented more than one element it was possible to ensure that the group was representative of most parts.

The stated task, as communicated to the membership, was concerned with "the identification of the following issues: what stress is occurring; where it is occurring; and, why it is occurring". The role of the consultants who worked with the members in large and small groups during the event was to provide sufficient containment to—as far as possible—enable the members to free associate to their experiences and feelings about their organisation. Another way of viewing it is the provision of a temporary holding environment that is experienced as "good enough", and develops the "basic trust" necessary for the members to freely communicate their feelings about the organisation.

As with individuals, who may be well-defended, it is also the case with the collective membership of an organisation, as it was in this instance. That is, collectively, it was exceedingly difficult for them to disclose much specific information at all, and they were very guarded in the information they did disclose. They were basically only prepared to work at a general level and when asked to be more specific they drew back, and disengaged from the task. In one sense this was experienced as not being helpful, but at another level it provided some important information about the way that they related to the organisation in their minds. Not surprisingly, the consultants experienced themselves as not being trusted, which provided them with transference information about how the members viewed the organisation. However, while lacking in specifics, over the 2 days the members did generate a great deal of data about the organisation, which they were then (as in Listening Posts), encouraged to consolidate into some form of hypotheses or statements. These were as follows:

Communications: Because of ineffective, inconsistent, or unnecessary communication from the centre outwards, along with a lack of trust in communication, stress is caused for everyone, but especially middle managers. This is manifested in lack of trust, distress, anxiety, anger, frustration apathy, "us and them" feelings, cover-your-back mentality, loss of self-esteem, tension, and a lack of belief that anything critical goes back to the centre.

Change: Because of the imposition of change allied to the pretence that it is always good, and the lack of recognition about the resulting increasing workload, stress is caused for all staff, but especially staff with a duel manager/care role. This is manifested in resistance, sabotage, uninvolved powerlessness, pressure, demotivation, insecurity for the future, team breakdown and stress when staff are ill.

Management Style: Because the management style from the top downwards and across the organisation is inconsistent, this leads to a lack of mutual respect, which in turn causes stress for all staff, but especially managers and senior managers, and staff at the point of service delivery. This is manifested in a high level of sickness, fear, friction, frustration, demotivation, low morale, and resentment.

Contracts and Finance: Because of lack of consultation and two-way feedback between senior managers involved in the contracting process and service deliverers there is inadequate planning and resource management. As a result, stress is caused for middle managers and service deliverers. This is manifested in work overload, reduction in quality of patient care, patients suffering, frustration, burnout, reduced productivity, and the staff feeling undervalued.

Shared Vision: Because of lack of a common vision, all staff do not work to common values or sometimes demonstrate conflicting philosophies and approaches. This causes stress for all staff, especially between localities and directorates, which is manifested in: anger; feeling undervalued and ignored, demotivation, lack of commitment, lack of productivity, lack of communication, reduced quality of services, avoidance, lack of loyalty, low self-esteem, lack of direction, staff turnover, absence, wasteful tendering exercises, and irrational behaviour.

Given the above, we now had a fairly clear picture of the external holding environment—the physical and social part—which showed that the ruling coalition was experienced as a "bad" object, and that lack of clear vision, policies about change, and internal structures all contributed to this view. Another issue, which was not quite so clear, but one which was beginning to come to the fore, was that concerning anxiety about clinicians who were now also expected to take on the role of managers as well as continuing in their clinical roles.

It was also clear that environmental influences regarding change were impinging on the organisation. Most obvious was that concerning the purchaser-provider arrangement where the health functions were now divided into two parts; the purchasers of health services, responsible for assessing the health-care needs of the local population and buying services, and the providers of health services such as hospitals. There was also some indication that the change from a national to a more local health service was also an influence. Both of these issues, which had been strongly opposed by health-service employees, were centrally inspired, government initiatives.

From the analysis of the transference/counter-transference material—that is, the way that the consultants were used by the Listening Group members—we also had some information about the internal psychological holding environment. From these data it seemed reasonably clear that the members' experience of the organisation in the mind was as a frightening, negative, and untrusting object. However, we still needed to get beyond this to gain a deeper understanding of the internal holding environment that we were progressively working towards.

The opportunity for advancing our understanding came at the end of the Listening Group when the members of the project management group were invited along to the last session where the members reported their findings. The structure of this meeting was as an intergroup event. The largely unconscious processes of the event resulted in the Listening Group members coming to represent the members of the organisation, and the project-management group being cast in the role of senior management. It proved to be a highly emotional session, and an extremely uncomfortable experience for the project management group. However, it was also exceptionally revealing in providing data about the members' views of the organisation in the mind.

In general terms, members expressed a great deal of passivity, resentment, and denial of their own creativity, and they tended to speak for themselves, or for their own particular part of the organisation, as if they were not part of the totality. In addition, they strongly expressed their lack of "basic trust" in the holding environment when they expressed the following with a great deal of feeling: all of the findings would not be carried forward for consideration; key issues had not been fully identified; they would not be involved in the follow-up; and they did not trust the project management group to address the issues raised. As their lack of basic trust had been developed as a result of their experience of the organisation holding environment, we can assume these concerns were to a greater or lesser extent based in reality.

These issues were further explored at a subsequent review meeting with the project-management group who expressed the view that they felt as if they had been attacked by the members of the Listening Group and that they had not been trusted to carry out the work. The following are some of the comments of the project-management group: "It is clear that members would not 'name names'"; there was a belief that "Things got stitched up behind the scenes, therefore it was not worth being a member of the group"; "People were scared to say what they think"; "It's like the secret service, MI5"; "It was as though they (the Listening Group) wanted the management group to be helpless, they were resistant and didn't want anything done"; "They felt that it was really only sham consultation"; and "There was a monster in the senior management team that couldn't be named".

By this stage, the intervention process had provided more than enough data for an analysis of the main features of the organisational culture. The external or physical and social part of the holding environment that the members of the organisation were interrelating with was experienced in a negative manner. It seemed that issues of lack of trust and openness arising from a situation described by the members as their experience of things being done "behind closed doors" or "through the backdoor", in a "secret way", resulted in members of the organisation feeling that "people are scared to say what they think", and adopting behaviour that is about "not naming names" and "defending their own corner".

In addition, however, in much the same way that unconscious forces operate in the maternal holding environment so too were they at play here in the organisational holding environment. This was experienced in a similar way to that described by Bion (1963). Here he observed that if the mother does not effectively help the baby by providing suitable containment the baby tends to introject not a comforting progressive experience but what Bion referred to as a "severe contentless anxiety". This, then, was the nature of the internal or psychological part of the holding environment.

With the hindsight of this new and deeper understanding, it was not difficult to see what had been blocking senior management and others from bringing about change and why the chief executive had so emphatically stated, "I can't unlock it". As a result of their interrelatedness with the organisational holding environment the members of the organisation adopted those forms of behaviour—which included the passivity, resentment, and denial of their own creativity—that they felt were appropriate to them under the circumstances that they perceived were imposed upon them by their organisational holding environment. Viewing the organisational holding environment, or the organisation in the mind as outlined above it is not surprising that the forms of behaviour adopted by the members of the organisation should be largely anti-task.

However, while these data in themselves may be informative, they are most valuable when we seek to make use of them to bring about change. This was the goal of the next stage of the intervention when we ran a 2-day workshop for some 30 members of the organisation, including all but one of the senior management team, and middle managers and others who were felt able to contribute to the task. The task as outlined was "for participants to develop their understanding of the structures and dynamics which are generating stress, and to apply that understanding to improving working relationships within the organisation".

Influenced by the knowledge of the organisational culture, we felt that by inviting the membership to view things indirectly that a transitional space would be created for them to constructively work at the task without going into some form of denial. We decided, therefore, that one of the main events should be based on a role-play exercise. After an initial period working with and gaining acceptance for the statements produced by the Listening Group, the members were invited to role-play four groups: senior management, middle management, mental-health practitioners, and community practitioners. The result was quite dramatic and resulted in sufficient confidence and trust for "names to be named".

Members felt free to name the individual who had previously been described as "the monster in the senior-management team that couldn't be named". By a mixture of good fortune and good planning we were able to use this information in a most effective manner. Because the senior-management team was present, this allowed us to start to work with the data. It became clear that there was splitting within the senior-management team. The senior manager responsible for service delivery was being asked to play a role on behalf of the other senior managers. They were using him by splitting off and projecting all their incompetence and hateful feelings into him as a defence against the anxiety they were experiencing. All of the loving and positive feelings were being split off and projected into the person that "got things done behind the scenes"; the "nice" chief executive, who was the "good" object for them.

The effect of this splitting was, on the one hand, to create some sort of monster who was hated and was to be avoided at all costs. On the other hand, it also resulted in the idealised "nice" chief executive who could do no harm. In turn, this had a detrimental effect on authority relations within the organisation, because, things "done behind the scenes" undermined the authority of other senior managers. This resulted in all senior managers being perceived as ineffective by the members of the organisation, and this contributed greatly to the organisation in the mind being viewed in a negative manner.

CONCLUSION

As I have tried to show, the way this organisation was perceived by its members, resulted in a negative organisation in the mind, which affected the organisational culture: "the way things are done around here". This is a useful piece of data, but leaving it at that level does not provide a strategy and means for how to influence the way things are done. It is only when we start to look at things from the perspective that "the way things are done around here" will depend upon the way that the holding environment is perceived, that we can begin to get a clearer picture. That is, we cannot influence culture per se but we can influence the organisational holding environment, and through this begin to modify the organisational culture.

In the sort of situation described above, where the organisational culture is influenced by considerable anxiety resulting in social defences, it may be a relatively slow and time-consuming process to bring about desired change. To get to a position where there is sufficient "basic trust" in the holding environment will not be easy. As Winnicott (1988) noted when referring to basic trust, "The word trust in this context shows an understanding of what I mean by the building up of confidence based on experience, at the time of maximum dependence, before the employment and enjoyment of separation and independence" (p. 120). In a similar way basic trust is developed as a result of the members' perceived experience of the organisational holding environment

Nevertheless, an awareness of their current organisational culture should provide management with opportunities to create the sort of formal structures, strategies, policies, style of management, values, and attitudes that will reflect an open, trusting, and reliable holding environment that is viewed positively by the members of the organisation. One where, for example: decisions are seen to be made openly and justly; there are no behind-the-doors deals; there is no secrecy; and one where co-operation as opposed to defending one's corner becomes the norm. When their concept of the organisation holding environment—or their organisation in the mind—is seen as a good object, the members of the organisation will begin to adopt progressive forms of behaviour that are task related.

By their actions and omissions, senior managers are, in general terms (like the mother in the maternal holding environment), key players in the organisational holding environment. As has been stated, the physical and social part is particularly influenced by factors such as the leader or ruling coalition, and the policies, structure, and strategies that they pursue. This in turn influences the psychological part, which is largely unconscious and is the basic social character of individuals and groups within the organisation and is based both on their current and past experience. It is important, therefore, that senior managers must not only understand in a deep way the nature of the organisation's culture, but the ways in which they themselves are influenced by it, and as a result, the ways that they may influence or perpetuate it, even if (and usually) they do so inadvertently.

This really goes to the heart of the need for understanding organisational culture. Large-scale interventions—and possibly others—must always, and no matter what else and with whom they are aimed at, work in a focused way with senior management in relation to the issue of culture, and the kind of holding environment they have created. Therefore, any large-scale organisational change must at least begin with senior management changing itself, and in doing so, beginning to create a qualitatively different holding environment that in turn will, as stated above, then drive other cultural changes in the organisation. Having an understanding of the way that culture develops will be a necessary and helpful starting point.

From the foregoing it will be appreciated that there can be no blueprint for changing culture, since every organisation and every culture is unique, and application of theory will result in different findings in each organisation. However, knowing how culture develops and the impact that it has on organisations allows us to apply the theory in any circumstances, thus providing us with the means to analyse and influence this complicated phenomenon.

REFERENCES

Bion, W. R. (1963). *Elements of Psycho-analysis*. London: Heinemann.
Dartington, T. (1997). Listening posts and the millennium. Paper presented at OPUS, London.
Jaques, E. (1955). Social systems as a defence against persecutory and depressive anxiety. In *New Directions in Psycho-analysis*, ed. M. Klein, P. Heimann, and P. E. Money-Kyrle, pp. 478–498. London: Tavistock.
Kegan, R. (1982). *The Evolving Self*. London: Harvard University Press.
Khaleelee, O., and Miller, E. J. (1985). Beyond the small group: society as an intelligible field of study. In *Bion and Group Psychotherapy*, ed. M. Pines, pp. 354–385. London: Tavistock/Routledge.
Menzies, I.E.P. (1960). A case study in the functioning of social systems as a defence against anxiety. *Human Relations* 13:95–121.
————— (1988). *Containing Anxiety in Institutions: Selected Essays Vol.1*. London: Free Association Books, pp. 43–88.
Miller, E. J. (1989). *The Leicester Model: Experiential Study of Group and Organisational Processes*. London: The Tavistock Institute.

Miller, E. J., and Rice, A. K. (1967). *Systems of Organisation: Task and Sentient Systems and Their Boundary Control*. London: Tavistock.

Rice, A. K. (1965). *Learning for Leadership*. London: Tavistock.

Stapley, L. F. (1996). *The Personality of the Organisation: A Psycho-dynamic Explanation of Culture and Change*. London: Free Association Books.

Trist, E. (1990). Culture as a psycho-social process. In *The Social Engagement of Social Science*, ed. E. Trist and H. Murray. London: Free Association Books.

Turquet, P. M. (1974). Leadership: the individual and the group. In *Analysis of Groups*, ed. G. S. Gibbard, J. J. Hartman, and R. D. Mann, pp. 337–371. San Francisco, CA: Jossey-Bass.

Winnicott, D. W. (1971). *Playing and Reality*. Harmondsworth: Penguin.

——— (1988). *Human Nature*. London: Free Association Books.

8

Institutional Learning as Chief Executive

EDWARD R. SHAPIRO

From the experience of directing a small program in a large institution (McLean Hospital in Boston), I was invited to lead a small, but distinguished, psychiatric institution, the Austen Riggs Center, as its medical director and chief executive officer (CEO). Immersed in Tavistock and A. K. Rice Institute Group Relations Conferences as well as organizational consultation for almost two decades, I felt this was the opportunity to apply what I had learned. In retrospect, the concepts that I found useful included:

- The distinctions between power and authority;
- The centrality of an institutional task as link to a shared reality;
- Authority as derived from that task;
- Roles as functions of the task;
- Listening to participants in terms of, "How are they right?";
- The learning in the here-and-now of organizational life;
- Attention to symbolic communication about shared unconscious institutional functioning;
- Leadership as a function of the system;
- Delegation to frontline roles;

I would like to thank Joy Bonnivier, Diane Jackman, and Drs. Virginia Demos, Donna Elmendorf, M. Gerard Fromm, Eric Plakun, John Muller, James Sacksteder, and Ess White for their helpful comments on an earlier version of this paper.

- The use of differentiated voices in subgroup tasks; and
- The importance of discovering links to the larger society.

These concepts derive from a framework of Tavistock group relations work that focuses on open systems notions. These include *primary task, boundaries, large- and small-group processes, institutional dynamics and culture, consultation*, and the importance of *interpreting experience in role* (Shapiro and Carr 1991) as a way to grasp essential aspects of organizational life. As I entered the Austen Riggs Center, I found myself searching for these familiar landmarks in this unfamiliar culture, and I subsequently grafted the organization's language onto this frame. Developing a new language with the staff and the board required extensive, disorienting—and often painful—negotiation. With some difficulty, *institutional mission* substituted for *primary task; personal boundaries* (the familiar language of psychoanalysis) expanded to *role; subsystem, and institutional boundaries;* the study of *large- and small-group processes* followed a public reassessment of the use of groups in the institution; a shared recognition of *institutional dynamics* emerged from the gradual development of a culture of negotiated interpretation;[1] and I adopted the use of *consultation* through *interpreting experience in role* as my management and leadership style. The central dilemma in this institution was how to maintain its link to a valued tradition at the same time as it joined a changing world. To help the staff accomplish this, I had to introduce the unfamiliar, while claiming the past. These linking concepts provided the necessary frame.

When I first arrived, the world of American mental health was in chaos as a consequence of the managed care revolution that had suddenly and powerfully constrained the finances available for health care in America. As a long-term treatment institution, the Austen Riggs Center was significantly affected. When I met the staff for the first time, I told them the following story:

> The world is going to end in a flood in ten days and the heads of the major religions get up on international television to speak to their flocks.

---

1. In any system, individuals in different roles that relate to a common task can have varying ways of making sense of their experiences. After agreement on the relevant context for an interpretation, individuals can arrive at a shared picture, if each person makes an effort to link his experience in role to that of others. Thinking "How is the other person right?" can lead to a negotiated, collective interpretation of organizational experiences (Shapiro and Carr 1991).

The Pope is first and says, "We have to repent and we will all join the kingdom of heaven." The Buddhist leader is second, and says, "We have to join the eternal oneness, and we will transcend this disaster." The Chief Rabbi is last. He says, "We have ten days to learn how to breathe under water."

This was 1991. As I was considering the move, the field of psychiatry was under major stress from the larger society. Accusations of plagiarism, nepotism, misappropriation of funds, and sexual misconduct had forced department chairs of psychiatry to resign. Psychotherapists were causing scandals about the loss of professional roles with patients, court cases were emerging about the implantation of false memories of abuse, and long-term treatment was under close scrutiny. Fiscal pressures were changing the landscape, and psychoanalytic treatment was under siege from biology and behavioral science.

The Austen Riggs Center was a distinguished, not-for-profit psychiatric institution whose staff had, over the years, made major contributions to psychoanalysis and ego psychology. It had represented the best in psychoanalytic theory and practice. Giants in the field, such as David Rapaport, Erik Erikson, Margaret Brenman, and Roy Schafer, had made significant contributions while at the Center. Riggs was a reservoir of psychoanalytic thinking and one of the last psychiatric centers that provided intensive psychoanalytic psychotherapy for disturbed patients in a long-term hospital setting.

As an institution, it was in major difficulty. Under the same pressures as other psychiatric institutions, it had more than a million dollars in accounts receivable in the context of a 6-million-dollar budget, with a small endowment. In the face of the fiscal pressures, the board was in a struggle with the medical director, and was suspicious of the private practices of the hospital staff. Morale was poor; people had been laid off. The medical director had resigned and agreed to stay on in a new office for an interim year while the search was under way. Riggs was an organization relatively isolated from current changes in the field; it had its private way of working. As with many psychiatric institutions of the period, the Center was organized as a straight hierarchy, with the medical director making all major decisions.

There was no human resource function, and consequently no established institution-wide personnel policies or procedures. The administration of benefits often fell to the department manager's interpretation for specific individuals. Job descriptions listed a string of tasks but no measurements or expectations. Performance evaluations were missing. There was no differentiated compensation system and everyone expected a salary adjustment every year no matter what their performance. The place ran on oral tradition. When I asked why something was done, the usual answer was, "Because that's the way we've always done it!" Managers were not trained in management. In many cases, in response to newly developing external accrediting pressures, brilliant clinicians had taken up quasi-administrative roles which they experienced as "tacked-on responsibilities." Patient care was their priority and they handled employee issues concretely without grasping the impact on the wider system, or the implications for patient care. The staff hardly discussed or attempted to understand what was going on in the larger system. Individual psychotherapy was the dominant discipline; there was one full-time social worker. Nursing staff felt devalued and therapists used them as supports for the psychotherapy.

POWER, AUTHORITY, AND TASK

The dynamics of the institution indicated a systematic confusion between power and authority. *Authority* derives from a shared task, and members of different subgroups in the institution did not seem in agreement about the task. *Power* relates to the availability and deployment of resources, and is either task related or not. My concern was that in a system that used power without a clear connection to a task, individuals can feel abused. This experience is often accompanied by splitting and projection, and that is what I was seeing between the board and the staff. I had learned that a shared and agreed-upon task can serve as an abstract "third party" that allows members to bridge their polarized connections and grasp a shared reality (Shapiro and Carr 1991). This system was under stress with marginal resources. I felt that its members needed to authorize an outsider to represent and help articulate the task on behalf of the system and help both sides find their way. They needed a consultant. In the year prior to my arrival, the board of trustees and the medical director invited me to take up this role.

I realized that not only did I not know what institution I was joining, the staff and the board did not know what medical director they were getting. From Boston, I set up a structure to negotiate a mission. I invited patients, staff, and the board to organize subgroups for the purpose of articulating their distinctive views of the Center's mission. I asked each subgroup to authorize representatives to negotiate the differences. Finally, I asked the entire system to authorize representatives to negotiate a final version of the mission with me. The process revitalized us all. Each sub-group engaged in a series of lively discussions that helped rediscover their values and beliefs, and their ownership of the institution and its traditions. Each articulated distinctive but overlapping areas. The final negotiation allowed me to grasp the institution and discover the connections I was bringing in from the outside.

Much of what I added linked the institution's values to the realities of the outer world: families, society, and the changing external context of managed care. The final statement was complex, but clearly articulated both the Center's traditions and some hopes I had about its links to the outside. The process deeply authorized me to take up the role of medical director/ CEO. The negotiated mission became the context for our work over the first three years.

How Are They Right?: Learning in the Here-and-Now

When I arrived, I was immediately immersed in issues of power and authority with staff and patients. The problem was to discover a shared context. The patients' reactions were symbolized in the following story:

> In negotiating the details of the medical director's residence, I had inquired about a piano. The retiring medical director had decided with the trustees that the patients were not using all of their five pianos and that one of them could easily be located in the new medical director's house and used for ceremonial occasions. The hospital community was in some disarray around the management of resources and power, given the transition in leadership. One consequence was that no effective discussion about the piano was carried out with the patients prior to my arrival.
>
> When I arrived for my new job, the patients greeted me with outrage that I had "stolen" their piano. Even though they were not using the piano, it "belonged" to them. In their experience, I was the CEO with all my perks and they were the abused victims of forceful power. As I saw it, some sort of negotiation had occurred before I arrived and they were reacting powerfully in a way I could not grasp.
>
> So the patients and I met—40 of them and me. We attempted to negotiate a shared reality, with frustration on all sides. The discussion focused on power—who controlled the pianos, they or me? There was

no possibility of neutral ground. On the face of it, the question seemed perplexing, since pianos were a resource of the institution and I was in charge of resources. Asserting that, however, would have been a power operation, since I could not at the moment discover the task that the resources needed to join. Since I was not about to act without understanding what we were involved in, we were stuck. I tried to listen to how they were right, but could not find the appropriate context.

But then one patient spoke movingly of the terrible sense of helplessness she had felt when the piano was arbitrarily moved without her consent. Though she did not play the piano, she felt strongly that something terribly important had been taken away. With a barely perceptible shift, we suddenly found ourselves talking about money, insurance, third-party payers, and managed care. The patients had entered the hospital and begun their engagement in treatment. Suddenly, without their participation, their financial resources were ripped away, arbitrarily, and irrevocably. The piano had suddenly become less important. We had discovered a larger context for this discussion; there was "a third." We were now talking about the task of treatment and the resources for providing it. I could join them, not by projecting negative images about power into managed care companies, but by working with them on the feelings of helplessness and vulnerability they had in their patient role about the encroachments of reality and limited resources. These were feelings I also had in my role as medical director. In fact, some of these feelings had contributed to my anxious wish to provide a formal space in my home to bring in outsiders and raise money for the hospital.

When we returned to the piano, the patients and I found that we could negotiate a process for its review, discussion, and decision. We had found a context for negotiating an interpretation of reality: the shared task of treatment we were all engaged in through our various roles.

But how could I be sure that my interest in the metaphor of limited resources and the apparently negotiated interpretation of the third-party payers was not simply self-serving and designed to allow me to displace my own arbitrariness and rigidity and facilitate my keeping the piano? Given that the patients and I together represented a system in enormous flux, both inside and outside of the hospital, it may have been too much to expect that they and I alone could hold to the task of treatment long enough to negotiate a view of reality without the rest of the system. Though the discussion continues, the piano is still in my home. The focus has shifted to other issues.

Perhaps one of the functions of a discovered context is to help with the integration of what might seem on the surface as competing experiences. The patients were regressively experiencing a repetition of unempathic, arbitrary power; I was in a similar regression, feeling misunderstood by them. In this mutual experience of empathic failure, both sides felt hurt, abused, and unable to learn from each other. Our discovery of the shared treatment task allowed us all to recognize our connections, recover from the mutual regression, and join in an interpretation of a shared reality.

The evidence that we had found at least the beginnings of such a negotiation came 5 months later. The patients left me a Christmas stocking on my office door that contained a beginning integration of ambivalence. Inside the stocking were two offerings: a lump of coal and a beautiful tiny piano with a tag that said, "This one's on us!"

SYMBOLIC COMMUNICATION

Senior medical staff sat together at lunch. Shortly after I arrived as CEO, I joined them, taking the seat at the head of the table. As I sat, there was a sharp intake of breath and several said to me at once: "You can't sit there!" I got up. Members informed me that the seat was reserved for the most senior clinician in the institution (a former director of psychotherapy); it was *his* seat. This turned out to be a sentinel event, alerting me that in order to take up my role I had to attend to the power of tradition in the institution.

I took this very seriously. This clinician was one of three partially retired senior staff. They did not talk with each other and were never in the same room together. Staff members assumed a deep disconnection between them and found it painful. Not attending to institutional meaning, staff gossip attributed the disconnection to conflicting personalities. All three were held in awe by the clinical staff. As icons, they represented the Center's great past. I invited the three for dinner. Staff members were shocked at my temerity and developed fantasies about the fireworks that would result. In fact, the three seniors were grateful for the invitation. As we socialized over dinner, their institutional roles emerged. One represented the link between Riggs and Chestnut Lodge (another prestigious psychoanalytic institution), and the centrality of intensive individual psychotherapy. The second was the former director of psychotherapy, representing the therapeutic community and the open hospital. The third represented a connection to Erik Erikson and research at Riggs. I began to wonder about the institutional reasons for their disconnection. As I formulated this question, they began to tell stories about these roles, leading to a warm and engaged evening. I articulated the aspects of the institution I heard from them and invited them to represent these within the institution. They seemed delighted. Their agreement to work with me meant symbolically that the new administration had respectfully engaged the Center's tradition. Over the last 6 years, each has become gradually involved. The first began attending case conferences on a regular basis. The second became director of alumni projects, organizing former patients and staff in a developing association. The third became a research affiliate and joined the staff for four months as Erikson Scholar. Six years after my arrival, Riggs hosted a gala celebration of the senior clinician's seventy-fifth birthday. I presented him with an engraved Riggs chair to place at the head of the senior staff lunch table, where he continues to sit today.

THE ORGANIZATION

The Austen Riggs Center had many brilliant clinicians and a creative clinical structure. It housed 50 patients in one program: long-term hospital care. The hospital was completely open, with no privilege system, no seclusion rooms, and no restrictions. Patients were free to come and go; many were severely disturbed. The Center had a therapeutic community in which patients were invited to take charge of their lives and contribute to the functioning of the hospital. There was a unique activities program started by Joan Erikson, where patients temporarily left the patient role and worked as students with craftspeople and artists. The intensive psychotherapy was well-organized and supervised, with a clear psychoanalytic tradition and boundaries.

The major problem was in the organizational structure. The world was changing rapidly, and the institution was moving inflexibly like an ocean liner. Changing direction took too long, and time was a limited resource. We needed a speedboat. Given the financial shape of the Center, I could not bring outside staff with me. My first task was to join the institution and attend to its internal world. I had to negotiate a way of working that would allow us to be more flexible.

The existing organizational structure had the medical director and administrator leading a hierarchical organization. Without a shared picture of the institution and its task, each department constituted its own world; the organizational atmosphere was filled with competition and rivalry. Information was a means of gaining power and control. One staff member said, "It felt like we were in the same boat, but each rowing in different directions."

Leadership As a Function of the System

Staff immediately turned to me for decisions for which I had little or no information. They asked me to approve funding for various projects, but they had no data about comparative budgetary needs, priorities, or benefits to the institution. Delegation of authority and flattening out the hierarchy seemed essential, if anything was to get done. I invited staff members to learn about the rest of the institution and take authority for decisions in their areas, from housekeeping to clinical care. We began the process of negotiating what I needed from them and learning what they needed from me. This proved slow and painful. Staff did not know me, and weren't sure they wanted to take up any authority. It was a dependent culture. Though staff took authority as individuals for the clinical work, their links to a shared institutional authority were difficult to discern; I had to be consulted on everything.

I broadened the executive committee and encouraged its members to notice my ignorance and interpret blind spots both in my understanding of the institution and in my style of working. I pointed out that, in my experience, significant problems in institutions were caused by such blind areas in leaders. Unaddressed, these could be experienced by the staff as power operations, separated from issues of task. I told the executive staff that I would speak to them about areas they might not notice, and hoped they would work with me in a similar way. My invitation seemed unsettling. When several dared to confront me and I was able to learn publicly from their observations, staff members began to take up their own authority. They began to see that I believed that *leadership* belonged to the system. I articulated my view that the person who most clearly could discover and articulate the task in any particular situation became the leader. This was an unusual authorization and was felt deeply by staff members. One (John Muller) noted, "Few executive gestures cost so little and have such immense impact as recognizing task leadership in employees."

It was difficult to grasp the way the clinical operation worked. The only role that seemed valued was that of individual psychotherapy. One full-time and one part-time social worker responded to social work needs for 50 patients, and nurses ran a separate culture in the patient building. Aside from the administrator and various credentialing committees, there was little formal administration other than a few business office personnel. The nursing director (Joy Bonivier) noted:

> When I had a problem in the past, I would take it to the medical director and expect him to "fix it." I did not perceive my role authority as including the responsibility to not only identify problems, but propose solutions. This dependency filtered down to the staff level with members complaining to me about issues but assuming no responsibility for addressing them directly. Staff would complain about a co-worker, but when I took disciplinary action, they would rally to the staff member's side in sympathy.

Delegation

There was a diffuse blurring of roles between nurses and mental-health workers, psychiatrists and psychologists. Homogenization of staff avoided feelings of competition and inequality, and resulted in dilution of role authority. Staff could not be assigned responsibilities based on differentiated abilities and training, so delegation was limited. Staff in leadership roles tended to "do it all" themselves, both because delegation was not accepted practice, and because doing it all gave the illusion of being indispensable. This process contributed to stagnation in professional growth, fragmentation in operations, and a crippling of role authority.

For example, department managers were not authorized to develop their own budgets. Instead, the administrator would present each budget based on historical data, "with a little added for inflation." Consequently, managers felt neither in charge nor invested in finding ways of saving money. The problem was in managing a shift from dependency in a hierarchical power structure to interdependency and team functioning in a system with a shared mission.

I delegated to senior staff members unfamiliar directorial responsibilities: clinical care, education, the community program. I authorized the director of admissions to manage the external boundary around managed care and the changing health-care world, and asked him to bring his learning into the system by taking charge of program development. I hired human resources and marketing directors. I sent administrative staff to group relations conferences. As staff members took up these new roles, they began to negotiate with each other, feel the shape of the institution, and see the needs of the staff below them. Grasping a changing administrative structure was a developmental move for many of the staff. They gradually recognized that good administration was a way to provide for the next generation. They discovered the generativity in creating the conditions for good clinical work and managing an open boundary to a changing society. They began to feel excitement at joining a larger sense of mission than they had previously grasped. From my perspective, I needed colleagues to help manage a complex system. Once they joined, we could together begin to discern the ways the institution was working and examine the place that Riggs occupied in the larger society. We gradually developed a work group to shape the institution to meet that need.

Differentiated Voices

Riggs' historical focus on individual psychotherapy and the professional power of medicine in a hierarchical organization had contributed to subduing the voices of other disciplines. Recognizing the importance of expanding an interpretive culture, we began an effort to strengthen these voices. Riggs had run on a model of large-group process for years. It had a morning conference, where all of the clinicians and nurses came to review all of the patients every day. People would say, "Morning conference decided this," and I would never know what they were talking about. The group seemed to organize itself through projections into the large group, where the voices of senior psychotherapists held all the power. Given that most patients had stayed at the Center 2 to 5 years, there was a shared sense of the collective, but I was stunned by the degree of irrationality that was contained (and promulgated) by the large-group process. The large group seemed to me to interfere with the delegation of authority and responsibility, so I attempted to shift the work into smaller groups. Against much resistance, we gradually developed four interdisciplinary teams to oversee the treatments of assigned patients. This marked a significant change in the direction of interdependency. Staff worried that this would disrupt the sense of "the whole," and the dependent connections held by the large group. To their surprise, however, staff found the change exciting. It was new to hear the voices of other disciplines discuss details of their work with patients. Patients began to attend team discussions so that they could join in thinking about their own treatment. Team leaders developed new administrative and leadership skills. The dynamics of the organization shifted from large- to small-group process: more coherent, differentiated, and graspable from the perspective of different roles.

Over the years, the Center had organized a creative therapeutic community structure around a series of patient government, support, and treatment groups. One patient-government group, "Sponsors," was responsible for managing dependency. In this group, patients assigned themselves to pair with new patients to help them learn about the institution. The group also took responsibility for tours of the Center for new patients and visitors. This placed the patients in charge of managing their living spaces and communicated to outsiders the values of the therapeutic community. Another patient-government group was called the Task Group. This focused on authority, and offered a structure for patients to examine community problems, particularly in terms of those patients whose behavior placed the therapeutic community and their own treatment in jeopardy.

There was a monthly patient-clinical staff meeting to review working relationships and a monthly All-Center Meeting. The All-Center Meeting is a large group consisting of all patients and staff (including support staff). The medical director chairs it with the patient chairperson (elected from the patient population every 6 weeks). The two of us meet to negotiate an agenda. Derived from our experiences in our roles, the agenda consists of our negotiated interpretation of the patient-staff boundary and our inferences and questions about the institution as a whole. Topics have included: Facing What We Can't Fix (about psychosis and limitations); What's in a Name (the use of titles and first names); Pressured by Our Context to Do What We Don't Want to Do (accreditation, families); Finding the Role of Citizen; and Interrelationships among Support Staff, Clinical Staff, and the Patient Community. This meeting is the one place where all staff members and patients can negotiate their understanding of the institution as a whole, incorporating the irrationality of the large group. Echoes of this meeting can be taken up in smaller groups and reconsidered over the ensuing month.

As part of our effort to shift to small group process and in response to requests from the patient community, we instituted "symptom-focused groups" on eating disorders, trauma, and substance abuse. We felt that symptoms were ways of managing meaning symbolically. We recognized how these symptoms, by inflicting meaning on others, evoked social reactions that isolated the symptomatic patients from the larger community.

The groups focused on the social communication inherent in symptoms, inviting patients to learn from this indirect communication and export their learning into the larger community of examined living.

In 1993, in response to a shift in the accreditation agency (Joint Committee on Accreditation of Healthcare Organizations), we re-articulated an aspect of the mission, focusing on the set of values that defined the institution (Shapiro 2001). This was an example of an external pressure that improved organizational functioning. Values are nodal points for staff and patient commitment and passion; articulating them allowed people more clearly to discover their connections to the mission. The values included:

- Affirmation of the dignity and responsibility of the individual;
- The importance of human relationships;
- Recognition and enhancement of individual strengths;
- Respect for individual differences;
- The learning opportunities in a community of differentiated voices;
- The importance of examined living;
- Attention to the conflict between individual choice and community need;
- Recognition and preservation of multiple roles, including student and citizen;
- Openness to innovation and creativity; and
- The centrality of the psychotherapeutic relationship.

The community agreed that without any of these, the Austen Riggs Center would not be recognizable.

THE LARGER SOCIETY

By the beginning of 1993, the census had gone up, and the financial picture had improved significantly. Then, President Clinton gave a speech to Congress about his plans to transform the health-care system. Within 30 days the number of admissions in East Coast psychiatric hospitals dropped dramatically. Private practitioners held onto their patients because they were afraid referrals would stop. The Austen Riggs Center began to lose a hundred thousand dollars a month. It became clear that attention to the internal world of the institution was not enough; we had to face our surrounding context.

We developed a small staff group that reassessed every job in the institution. We looked at what we absolutely needed to preserve the essence of the mission. The results led to a dramatic and painful downsizing, losing almost a third of the staff. This event was staggering for an institution where long-term staff had committed their lives and careers to a previously reliable culture. Though most of the departing staff members were able to find other positions, many remained in the small town of Stockbridge, where they passed the rest of us daily on the street. The pain of this decision continues to affect all of us.

The process deepened our interdependency with each other and the patients. As a consequence of the need to reduce staff, we placed more weight on the therapeutic community and on the patients' capacities to manage themselves. Authorization of patients' capacities was a long-standing strength of the open setting at Riggs. Deepening it allowed us to learn more about the ways our patients resisted self-authorization because of unconscious delegations from their families and society that were meant to be held, locked away, and not interpreted. Patients taught us how they had become "good citizens" of their families and social groups, by identifying with unspoken directives about the roles they were to assume. Unconscious compliance with these delegations interfered with their capacities to discover their own authority. Our authorization to patients to become citizens in our community of examined living encouraged their ability to translate these delegations (sometimes through their own enactments with us) and take up their own authorship of their lives.

We believed that the managed-care perspective was shortsighted. Mental illness is not a short-term problem, and we felt that the pressure for short-term solutions would not last. We recommitted ourselves to our mission and to intensive, four-times-a-week, psychodynamic psychotherapy in a treatment community that offered the longest opportunity to learn from this work. But we decided that the survival of this mission required taking seriously what the world was saying to us.

We approached this connection through families and the larger health-care world. Riggs had never paid much attention to the family, focusing on individual dynamics. When authorized by the patient, I began meeting with patients' families, and expanded the social work department to make a consistent link to families. The health-care world was saying that there were limited resources available. I organized a Resource Management Committee, bringing together businesspeople and clinicians to examine the management of limited resources and their clinical meaning. The task of that committee began to lead our institutional learning (Plakun 1996, Shapiro 1997).

INSTITUTIONAL LEARNING

As the Resource Management Committee began addressing the limited resources within each case, they ran into irrational responses to these limits from patients, families, and staff. Denial, rage, and projection were common on all sides. The presence of clinical people in these discussions required negotiating a shared language. We discovered that "limited resources" was both reality and metaphor. The reality required management. The metaphor, applied to financial, emotional, and family resources, required discovery and interpretation for each case. Staff and patients felt enraged about limitations, but focused that rage on managed-care companies, rather than integrating it into the treatment. The Resource Management Committee pulled together these pieces and began to work with patients at managing resources and discovering the appropriate metaphor within each treatment. These metaphors were much like a shared context. Connections initially denied through the operation of polarities could—through linking metaphor—be named and owned. For example, an adopted male patient's struggle with limited financial resources could be linked to his unconscious sense that his father had died too soon, depriving him of the necessary resources to become a man. Interpreting such a connection can transform paranoid and unworkable anger into manageable grief.

This new language both energized the institution and transformed it. Staff began to help patients take charge of the limitations of their resources, and understand the meaning of their reactions within their psychotherapy. The negotiation between business and clinical staff sensitized both groups to the others' world. Clinicians began to grasp the fiscal issues of families and managed-care companies. The learning helped the rest of us to reshape the Center's programs to meet these limitations.

With the patients' consultation, we developed a range of treatment settings which looked to managed care like six different programs: inpatient, residential, day treatment, aftercare, a halfway house, a residential apartment. Since Riggs was small enough for the same therapist and interdisciplinary team to follow the patient all the way through the spectrum, to those of us inside Riggs, it felt like one program. Given this development, we renegotiated our external boundary so that we could both maintain our mission and allow patients, families, and managed-care companies to save money. The spectrum concept made patients' living situations more flexible and reduced ancillary staff. The shift was largely in the minds of patients and staff, since two of these programs and the Community Center were in the same building. The collective work was to define a set of boundaries that would allow patients and families to assess—with our help—what length of treatment they needed, what behaviors they could manage, and what additional resources cost. With this information, they could select services so as to extend the length of their psychotherapy and manage the total cost of their treatment.

We also helped the board of trustees to reshape itself in order to bring people with specific expertise in the outside world: marketing, development, business. We expanded our Community Outreach Program. The Center's existence owed a great deal to the small town of Stockbridge that supported it. One of the patients said to us that Riggs was one of the best places she had ever been because it was on Main Street in town, not tucked away from sight in the mountains. We felt we needed to give something back. Building on the Riggs' tradition of helping develop the area's community mental-health system, we began consulting to local schools and not-for-profit organizations. We developed a Friends Organization of former patients, former staff, neighbors and interested laypeople. This helped us develop a network of supporters and colleagues who could help us relate to the outside world, improve our marketing, and support the future of the institution.

We began to articulate our voice. With the help of the board's marketing committee, we formulated a slogan: "The Austen Riggs Center: Where 'Treatment-Resistant' Patients Become People Taking Charge of their Lives." We put the phrase "treatment resistant" in quotes because our patients did not experience themselves that way; they experienced previous institutions as "treatment resistant." This continued the process of identifying the institution to ourselves and others.

Managed-care companies defined their outliers as "treatment resistant" and we turned to them. Riggs was receiving patients whose treatment had failed after up to one hundred short-term hospitalizations in other settings. We talked to companies about the fact that 10 percent of their patients—their "outliers," who did not benefit from short-term interventions—were spending up to 70 percent of their resources. We suggested that, in the long run, definitive treatment would save them money. In response, several companies developed contracts with us. During 1994, the same year as the downsizing, Riggs had its seventy-fifth anniversary. With extraordinary work from a traumatized staff, we held an international symposium, hosting 300 people from 11 countries and 28 states. Sixty former patients from 5 generations attended an unprecedented former patients reunion. They presented a fascinating panel discussion on their treatment experiences at the Center, one wearing a T-shirt saying "I survived Riggs." They reviewed the failures of staff in the 1960s to understand the nature of trauma, discussed the painful implications of working with therapy staff who followed rigorous psychoanalytic technique without seeing the individual "in context," and recounted the difficulties they had experienced in the abrupt transition from full inpatient status to the outside world. These events stimulated press coverage and our census grew.

The treatment world was telling us that we had to demonstrate our results. We organized a sophisticated study of treatment outcome, focusing on the differences between symptom change (which is rapid and temporary) and character change (achievable only with substantial treatment). Promoting this outcome research was a difficult problem: it was expensive, very sophisticated, and uncertain as to its outcome. I expended a significant amount of "leadership equity" in convincing the trustees to support it. I felt strongly that we needed to do more than assert the value of our treatment, we had to commit ourselves to proving it. We had to determine whether deeper treatment leads to less resource utilization. Coincident with our seventy-fifth anniversary, Plenum Press published an earlier Riggs's MacArthur Foundation-funded study on successful treatment outcome at Riggs (Blatt and Ford 1994).

These explorations helped us refocus our mission. We were seeing that the effective treatment of our patients required close attention to their contexts: families, the therapeutic community, the larger society. In fact, the external context was embedded in both of our earlier mission statements. In 1994, Erik Erikson died and we decided, with the help of his wife, Joan, who founded Riggs's Activities Program, and his children (one of whom, Kai, is on the board), to develop the Erik H. Erikson Institute for Education and Research at Riggs. Having spent over a decade on the Riggs's staff articulating the effect of context on individual development, Erikson was the perfect figure to symbolize our newly formulated mission: the Study and Treatment of the Individual in Psychosocial Context. The task of the Erikson Institute is to develop education and research at Riggs and to apply what we are learning clinically to larger social issues. With the formation of this program, we were institutionalizing our learning and educational boundary with the larger society.

More recently, we have begun to explore the following question: To what end are we treating our patients, beside the more general outcome of "mental health"? The half-century development at Riggs of a therapeutic community program had laid the groundwork for patients taking up community roles. We began to recognize the importance of that and see how "the study and treatment of the individual-in-context" was aimed at facilitating patients' reentry into society as citizens. Freud described the outcome of psychoanalysis as the ability to love and work. He never said anything about *voting* (Simon 1996)! Joining a world of equals in a fully participating role is a central aspect of psychological health. Our newly focused mission and our community of examined living allow us to work with these issues directly by focusing on the psychological components of competent social interaction.

The first Conference of the Erikson Institute in 1997 was on: "Psychotherapy in Changing Contexts: Losing and Finding Our Way." The design included formal presentations, small-group workshops, and process consultation. The process allowed us to learn about the intergenerational tensions in the field. The senior generation was experienced as having "answers" and filling up the space with them, while they experienced themselves as holding not only experience but endangered values. The middle generation recognizes the "old answers," but faces a new system where these answers need adaptation. They may be in the process of finding a middle way. The youngest generation was filled with chaos, change, and uncertainty, with no time for reflection. They have grown up in a new system, but find no time to discover their own answers, in part because the seniors pressure them with "old learning."

A group of the senior clinical staff of Riggs—all white men—organized a panel presentation in the conference. Four clinicians presented highly compact, interesting papers, but allowed no time for reactions from the audience. In the workshops, the participants saw this as a significant representation of Riggs. They helped us grasp the pressures on Riggs that could be leading to the development of a male, narcissistic institutional culture, where the staff had things to say, but allowed no responses from the outside. We were to be admired, not engaged. We had left ourselves no time to learn from the next generation, and we had failed to develop a senior staff of diversity.

We also learned about the isolation of psychotherapists who no longer can find reliable institutions that stand for the values of psychodynamic treatment. As a result, they cannot find ways to join with like-minded colleagues. Participants experienced the Center as an icon that held these values, generating powerful wishes in them for its survival in the face of a social transformation toward the biological and behavioral. Despite our flaws, they felt they needed us. They urgently encouraged us to diversify and open ourselves more to outside experience.

DISCUSSION

This chapter is a brief overview of significant change and institutional learning occurring in a small psychiatric organization in the shifting context of health care in the United States. It necessarily skips over the personal anguish, disrupted relationships, symbolic changes in personnel, and other more revealing complexity that would give greater texture to the story. It also presents the changes in a positive light—largely because I am writing with the narcissistic investment of a CEO. But the chapter would be incomplete without touching on the downside. This kind of sweeping change in any institution—and Riggs is similar to many changing institutions in contemporary America—comes with a price.

Over the years, we have developed both internal and external consultancy to the Center. Having invited staff members to attend to each others' blind spots and use their experience in role as an access point for institutional interpretation (Shapiro and Carr 1991), we had to listen to what we were saying to ourselves. We also invited psychoanalysts from Central Europe, who developed the practice of psychoanalysis under totalitarian communist regimes, to visit us and tell us what they saw from their outside perspective. The observations from inside and outside coincided. They were disconcerting.

Our European visitors noticed a kind of institutional complacency and narcissism. It escapes no member of the staff that we are in a privileged place at Riggs. The Center is one of the last psychiatric institutions in this country devoted to the intensive psychotherapeutic understanding of seriously disturbed patients. Our American colleagues in other institutions are struggling with these patients for 2- to 5-day stays, while our average length of treatment (in the range of settings) approaches 300 days. We are admired and envied from the outside, and run the risk of identifying with these projections and believing that we know what we are doing and don't have to learn from the next generation and the outside world. Our consultants advised us to cultivate internal dissent and differentiated voices.

The psychotherapy staff is now in three groups: senior full-time staff, fellows (we have a four-year fellowship for postdoctoral psychologists and psychiatrists), and a new group of part-time affiliate staff (one of the consequences of the downsizing). Because of external pressures for documentation and concreteness, we overuse our administrative staff and pull our senior clinicians away from direct patient care. As a result, there is a widening gulf between those who carry day-to-day clinical work and the administration. One consequence of that gulf is that the words of senior clinical staff are given too much weight and begin to sound like "the Party Line" (according to our Central European colleagues who have a great deal of experience with totalitarianism). A second consequence is that our part-time staff lose their connection to the therapeutic community and feel like "second-class citizens," without an adequate voice. With this evolving structure, we are running the risk of developing a group-based hierarchy and an institutional language characterized by pathological certainty (Shapiro 1982).

Our downsizing and efforts to empower patients have as a secondary consequence the attenuation of the supportive relationships between nurses and patients. With a larger staff, nurses could spend more time with individuals. Though we have tried to compensate for a smaller staff by providing a more highly staffed, intensive inpatient program, in the step-down settings there are patients who cannot be sufficiently held by peers and groups to take the risk of entering a deepening psychotherapy. Interrupting a dependent culture has consequences, if the resultant interdependency is not well grounded in the treatment task.

Finally, one of the unforseen consequences of flattening the hierarchy and encouraging role differentiation is a fragmentation of staff, with small work-groups pulling away to do their work. In contemporary America, the loss of a sense of community is a familiar phenomenon, with resultant atomization and isolation of the individual and subgroup. In Riggs's efforts to maintain its mission while meeting the outside world, have we joined that external fragmentation? Are we at risk of losing our Center? The cyclical nature of change requires vigilance to the possibility that while solving one set of problems, we generate another.

Nonetheless, the consistent search for the evolving organizational task that links staff and patients to each other and to the changing external environment continues to serve as a compass for grappling with these issues. It defines the identity of the institution, provides a matrix for collaborative learning, and serves as protection against the deadly narcissism that comes from isolation and disconnection.

I have found the past seven years in this role enormously stimulating. But the tension of trying to maintain Riggs' mission in the face of a rapidly changing society does not decrease:

> The story is told of a new chief executive who comes into an institution, and the departing leader gives him three envelopes labeled "1," "2," and "3." He tells him to open one at each crisis. The first crisis comes and he opens the first envelope and it says, "Blame your predecessor." The second crisis comes and he opens the second envelope and it says, "Blame the environment." The third crisis comes and he opens the third envelope and it says, "Make three envelopes."

REFERENCES

Blatt, S., and Ford, R. (1994). *Therapeutic Change: An Object Relations Approach.* New York: Plenum Press.

Plakun, E. (1996). Economic grand rounds: treatment of personality disorders in an era of limited resources. *Psychiatric Services* 47:128–130.

Shapiro, E. (2001). The changing role of the CEO. *Organizational and Social Dynamics* 1 (1).

Shapiro, E. R. (1982). On curiosity: intrapsychic and interpersonal boundary formation in family life. *International Journal of Family Psychiatry* 3:69–89.

——— (1997). The boundaries are changing: renegotiating the therapeutic frame. In *The Inner World in the Outer World: Psychoanalytic Perspectives,* ed. E. R. Shapiro. New Haven, CT: Yale University Press.

Shapiro, E. R., and Carr, A. W. (1991). *Lost in Familiar Places: Creating New Connections between the Individual and Society.* New Haven, CT: Yale University Press.

Simon, B. (1996). Can there be a psychoanalysis without a political analysis? In *Genocide, War and Human Survival: Essays for Robert J. Lifton,* ed. C. Strozier and M. Flynn, pp. 283–295. Lanham, MD: Rowman and Littlefield.

9

The Leader, the Unconscious, and the Management of the Organisation
ANTON OBHOLZER

INTRODUCTION

Leadership would be easy to achieve and manage if it weren't for the uncomfortable reality that without followership there could be no leadership except perhaps of a delusional sort. What is more, for the organisation to be creative it requires followership to be an active process of participation in the life of the common venture, encompassing all concerned, and this in itself carries with it a degree of discomfort.

By definition there is thus an inherent tension between leadership and followership. This chapter is an attempt to address the complexity of this interface, to place the relationship in the context of the overall containing organisation, and to investigate some of the factors that make for and facilitate a creative versus a stuck workforce and workplace. My overall approach is heavily influenced by a model of understanding and managing organisations that draws on an applied psychoanalytic and Tavistock group relations approach.

I want to note at this point that in many models of leadership and of management, working at understanding one's experience and the experience of others in connection with management and management competence do not necessarily go together. Further, that the sort of personal work and institutional introspection that goes with the approach to be described here is seen as unnecessary, gratuitous navel gazing. I believe this latter view to be profoundly shortsighted.

For the purpose of clarification I also want to draw on the helpful differentiation between organisation and institution made by my colleague, Wesley Carr (1996). He uses the term "organisation" to address the "bricks and mortar" components of the workplace and "institution" as the picture-cum-concept that we carry in our mind to describe the "philosophical" component. Armstrong's paper (1998) on the "institution in the mind" is also essential reading.

With regard to the organisation of this chapter, I begin with a discussion of the core functions of leadership and proceed, in turn, with discussions about related topics. I try to pull some of these issues together by concluding with a brief summary regarding the nature of leadership, followership, and the creative workplace.

THE CORE FUNCTIONS OF LEADERSHIP

The Primary Task and the Visionary Function of Leadership

Above all, leadership must be about a vision and a strategy for the future. I differentiate between vision and strategy because, in my view, there is and has to be a degree of passion and indeed of fervour in vision. Strategy by contrast is a "colder" element of leadership, but nevertheless an absolutely essential one in having a goal as to where the organisation needs to be in future years. In this sense, strategy also acts to temper vision and to make for reality in laying down achievable goals for the future. It is strategy that enables the vision to be achieved. But leadership vision unchecked by strategy and out of touch with an active work-group followership process, makes for the danger of a delusional system and, at worst, leads to the *folie-en-masse* seen, for example, in recent cult disasters.[1]

The main institutional "ballast" that keeps the organisation, both membership and leadership, steady must of necessity be the awareness of the *primary task* of the organisation. Miller (1967), Rice (1958), Obholzer (1994), and others have written consistently about the danger of the primary task being infiltrated and corrupted by defensive processes arising in

---

1. Jaques (1989) built a whole system of understanding institutions and leadership on the matter of time and the importance of leaders having the capacity to think ahead in substantially extended timeframes. Thus, the more important the task, the more the capacity of the leader to plan many years ahead had to be present. The leader was thus expected to work to a horizon that the followership generally were not occupied with.

response to the work of the organisation.[2] It is in my view one of the core elements of the task of the leadership of the organisation to see that the concept of the primary task of the organisation is not only uppermost in the minds of all the members of the organisation, but that it is constantly reviewed in the light of the external environment and that the functioning, structure, and staffing of the organisation changes in accordance with the changing primary task and its cluster of subtasks.

If there is to be an evaluation of whether members' work is of value to the organisation or not, then there has to be some baseline against which work-activity is measured. Without such a concept it is an institutional "free-for-all" as to whose view is to prevail, how one is to conduct oneself in one's work role, and how performance is to be judged. Later, Rice, working with Miller (1967), refined the concept, making the point:

> The primary task is essentially a heuristic concept, which allows us to explore the ordering of multiple activities (and of constituent systems of activity where these exist). It makes it possible to construct and compare different organisational models of an enterprise based on different definitions of its primary task; and to compare the organisations of different enterprises with the same or different primary task. The definition of the primary task determines the dominant import-conversion-export system. . . . . [p. 25]

However the primary task is defined, the important element of the concept in institutional functioning is the fact that a debate needs to continuously take place about what the institution is about, where it is heading, and that every "individual" contribution to the overall institution has to be held within group and institutional norms and boundaries. This concept is thus a key element in member/member and leadership/followership interaction.

Leadership and the Management of Change

From the above, it is clear that leadership is essentially about the management of change, both internal and external to the organisation, and the put-

---

2. Rice (1958) originally developed the idea of the primary task which he defined as follows: "Each system or sub-system has, however, at any given time, one task which may be defined as its primary task—the task which it is created to perform. . . . In making judgements about any organisation two questions have priority over all others: What is the primary task? How well is it performed?" (pp. 32–33).

ting into place and servicing of mechanisms that enable the two components to link and cooperate with each other at a pace of change that is emotionally possible and realistic to both external and internal needs. The key question here is whether change is ever "internally" driven, or whether it is inevitable that change is determined by changes in the external environment. With the possible exception of succession issues, most change seems to be driven by changes in the environment as it impinges on the workplace.

Change that has its origin in the environment cannot as it were be "wished" away and has to be taken account of, though obviously for a time the existence of the external pressure can be denied by the process of "turning a blind eye" to it (Steiner 1985). Change that is internally generated by contrast can be hived off or "encapsulated" (the institutional equivalent of the body walling off an infection or the psyche encapsulating a traumatic experience or part of the self). Unless there are external factors that can enter the institution and augment the pressures contained in the encapsulated part, it is likely that the process of institutional resistance to change reaches an equilibrium with resultant no change. Resistance to change therefore inevitably resides in the institution.

In other instances new ideas do arise in institutions, but what often happens is that the ideas and and those that embrace them form a discrete "enclave" within the organisation, and if they do find favour often do so by being taken up by others outside the organisation or by the creation of a new external structure to pursue the idea or produce the product. Leadership is therefore about managing a quite sensitive titration process—too much external reality overwhelms in-house values and the strengths of the past are lost, too little titration of reality and the organisation is at risk of being bogged down, irrelevant and eventually conceptually and financially bankrupt.

Leadership and the "Osmotic" Boundary-Keeping Function

For the work of the institution to be reality based, what needs to be aimed for is an institutional "chemistry" where there is sufficient awareness of the pressures impinging from and opportunities afforded by the external world. For example, a recent review of the British National Health Service caused a great deal of organisational turmoil as the system changed to a "market" orientation. The risk was for members of staff to either ignore the changes, thus putting the enterprise at risk, or alternatively for all members of staff to be so taken over by the "ins and outs'" of the market process that they neglected or ignored their history, traditions, and their core skills and contributions—in a word, both their institutional competence and wisdom.

What was required for sufficient awareness of change to enter the system was for key office bearers, through a process of "titration" of external reality to make clear that the core marketing responsibility would be undertaken by a small designated network of workers who performed this function on behalf of all. This thus left the rest of the staff to continue working at what they were good at but in a changed context as outlined above.

This "osmotic" boundary-keeping function is, of course, a two-way one, with the values, ideals, worth, and products of the organisation also needing to be communicated to the outside world. Further, in setting up and maintaining the necessary structure for the institution to run effectively, there also needs to be an awareness of everyday practical issues that are integral to the well-being of an organisation. Matters such as group size for effective working, clarity of task and role, and of boundaries differentiating various work-related functions (Obholzer and Roberts 1994). In order for this to be effectively managed and overseen, the leadership needs authority and power. Authority is a product of organisation and structure, be it external as in the organisation's sanction or internal as in the inner world of the leader or leaders' experience.

Leadership, Power, and Authority

However, authority, though necessary, is not sufficient. Power, which is having the resources to be able to enact and implement one's decisions is required as well. For example, a leader authorised by the organisation and personally in touch with his/her inner world issues to a resolved degree will, nevertheless, be quite ineffective if the means to effect decisions, money, staff, equipment, and so forth, are not available. In this regard it is also important for attention to be paid to the terminology of authority—appointing someone as chief executive gives quite a different message from appointing the same person as co-ordinator, even though they might have access to exactly the same resources to implement decisions arising from their authority base. All authority must be exercised in the context of the sanction of the followership as noted, but any such sanction or lack thereof must be measured against the benchmark of the primary task of the organisation.

By this I mean that the giving or withholding sanction for change must be measured against the benchmark of what the change is intended to achieve. If the change is in the service of developing and furthering the primary task of the organisation and approval is withheld by the follower-ship, then I believe that the withholding of approval should be taken to mean resistance to change. This would then require work on the part of the management to address the underlying anxieties that are presumably at the root of the process while also encouraging work to be done on the part of the membership to produce alternative models of achieving the necessary movement towards change.

In connection with the above, I believe an awareness of the presence and workings of unconscious personal, interpersonal, group, intergroup, and intrainstitutional processes among both leaders and followers to be essential. As a basic minimum such awareness can help to prevent one, in whatever role, from colluding with or being caught up in antitask institu-tional processes. At best such an awareness enables one to be proactive in expecting such processes to make their appearance at certain strategic stages of institutional development, and to ensure that they cause a minimum of disruption. Many of the unconscious dynamics of personal and family life are at risk of flaring up again and being enacted in institutional function-ing, particularly if the institution is led and managed in such a way as to turn a blind eye to such issues. This in itself can, for example, be a re-creation of a leader's personal past dilemma, say where a parent's ignoring the noxious effects of sibling rivalry, or worse, played upon these issues. Many an organisation is beset by a reenactment of such dynamics in their staff group, and whilst a "family type" or a therapeutic intervention is never justified, an awareness of the presence of such issues and resolutely not playing into them can go a long way in creating a task-orientated team.

Leadership and Institutional Dynamics

In many organisations thinking about institutional dynamics and their management is delegated to the human resources or personnel function. In my view this is effectively the equivalent of a splitting process with "wet" or "soft" elements of the whole, being disowned as being part of the whole and left to be dealt with (or more specifically, not dealt with) by only a part of the organisation, and an often derogated part at that (especially by line personnel). In therapeutic and human-service organisations the pro-cess often takes a related form—here personnel does not deal with it, but instead a member of staff with a suitable "valency" (Bion) for this sort of work takes on the role, often as part of an unconscious institutional process. The end result, however, is the same—it is not owned as part of the overall functioning of the system, and dealing with it "therapeutically" makes it unavailable for institutional inspection and work. This is because in the "therapeutic mode", be it individual or organisational therapy is private, personal, and confidential. A "therapeutic" style of management, such as is often found in mental-health-sector organisations, runs the risk of missing the primary task of the organisation, namely, the welfare of patients and instead replacing the task with a focus on the well-being of staff instead. Leadership and the management matrix should, of course, have a therapeutic and staff-development side to it, but in the context of the overall primary task and institutional processes.

In all organisations the health of the entire workforce must fall within the remit of management who also have a particular responsibility for minimising the effect of "toxic" processes arising from the nature of the work the organisation is engaged in. This applies equally whether the toxins are physical products inherent in the manufacturing process or whether they are toxins "in the mind" with an effect on the personality such as one finds in people changing organisations. Jaques (1955) and Isabel Menzies-Lyth (1990) were particularly innovative in this area when they wrote of social systems as a defence against anxiety. By this they meant that the anxiety arising from the nature of the underlying work and the breaching of societal taboos that are often an essential part of carrying out one's duties can have a very detrimental effect on the state of mind of the worker. As a result work patterns are organised in the service of psychic defence mechanisms, rather than in the pursuit of the primary task. It is axiomatic that the membership of the organisation will be particularly caught up in enacting such defensive processes, though personal valency (Bion 1961) and vulnerability, or lack thereof, to certain unconscious processes will obviously affect individual response. Bion borrowed the concept of valency from physics where it denotes the proclivity of an atom to combine with others. In his applied sense in the field of human unconscious group/institutional processes, he used it to mean "capacity for instantaneous, involuntary combinations of one individual with another for sharing and acting on a basic assumption". The connection between the individual and the institutional process thus being via the valency factor of the individual's personality. Additionally, a multidisciplinary or multigrouped organisation also allows for splitting processes to happen, so that the defences can take the form of intergroup and interdiscipline issues and rivalries, thus masking the defensive processes and the flight from work.

It is particularly tempting for those of therapeutic bent to intervene in a "therapeutic community" type of way, rather than to address the underlying issues; but then, even in a therapeutic community such interventions are, or were originally intended to bring out and address the underlying issues. By addressing them in a "therapeutic community" sort of way, I mean commenting on their existence or drawing them to attention, and possibly speculating about their origin and meaning without addressing issues of what's to be done to address them, by whom, in what role, with what authority and in what timescale. As mentioned by Mosse (1994) the problem that needs to be faced is that action taken by managers is often labelled by therapists as being a "manic response" while the "thinking" undertaken by therapists to address issues is often seen by managers as leading to endless speculation and no action being taken to address the problems. These views are obviously the polar ends of a splitting process—the goal to be achieved is thoughtful action.

The Core Functions of Leadership: A Brief Summary

In bringing this section to a close, it is perhaps helpful to summarise the core functions of leadership. Vision, managing the boundary "osmotic process" in the context of the strategic timetable and plan of the organisation, creating a containing structure conducive of creativity and thought, coupled with a relentless ferreting out of defensive, bureaucratic time-wasting activities in the service of resistance to change seem to me to be the essential requirements.

OTHER DIMENSIONS OF LEADERSHIP

Leadership and Management

If leadership has the qualities outlined above, what then of management, and how is it different? Management, if practised unimaginatively in my view, is leadership without the vision, and therefore to a degree the management and administration of the status quo. While this in itself is an important function, it is not enough to produce satisfaction in either the managers or the managed. The response then often is that the managers fall into increased states of bureaucracy, both "in the mind" and "in the system", and the managed fall into states of denigrating management. This can either take the form of casting managers as parasites who live off the work or creativity of the workers, or else for management not to be seen as real work. "When can we get this over with so that I can get back to my real work?" was the parting comment of a senior colleague who straddles a leadership/management working role. Not seeing management, and to a lesser extent leadership, as real work perpetuates the problem cycle and confirms the self-fulfilling prophecy about management. Doing the real work, whether it is as engineer, lawyer, or psychotherapist is of course also a "face-enhancing" or at least face-saving way out of coping with the difficulties of having to perform a managerial or leadership role.

Leadership Styles and the Group Process

At an unconscious level the leader is perceived as giving the group a message, namely "I think I'm better than you are". While leaders and followers would deny any such intent in their conscious perception and dealing with each other, I believe the facts speak otherwise. Leaders and managers are experienced as siblings who have reached "above their station" and in the cauldron of unconscious institutional processes are thus perceived as 'fair game' for a process of bringing them down to earth, that is unconsciously to the same "level" as oneself. This attack on a "sibling" is often dressed up as helping the individual(s) concerned to be more "in touch with reality"; help them therapeutically not to fall victim to "omnipotent projections"; or generally help them along. It would be wrong to see such activity only in terms of an envious attack on a sibling rival; it would be equally inappropriate to sweep these processes under the carpet to pretend that they don't exist.

Bion (1961) has described different states of group, and, I believe, of organisational functioning and thus of leadership. Dividing the most primitive mode of functioning, which he named "basic assumption functioning", he described basic assumption fight-flight, pairing and dependency. Each has an accompanying "in-house", "across-the-boundary", and "leadership" implication. The leader of an institution that is in fight-flight mode thus needs a fight-flight dynamic in order to function effectively, and, if that is no longer forthcoming on account of changed institutional or environmental conditions, is at risk of either gratuitously stoking up more fight or else of needing to change his or her style. There is of course no immutable basic assumption state, and there is always a "contribution" from other basic assumption modes to the overall functioning of the institution. Bion also elaborated the dynamics of a more mature style of institutional functioning which he described as the "work group". Here the direction of the institution is essentially aware of the primary task and the basic assumptions are harnessed in the service of carrying it out. Bion termed this process "the sophisticated use of basic assumptions."

Although Bion, in his writings, never made a direct link between his thoughts on institutional functioning and Melanie Klein's concepts of the paranoid/schizoid and depressive positions there is nevertheless a clear and helpful parallel between the two sets of concepts, with basic-assumption functioning and the paranoid/schizoid-position functioning having much in common, as do work-group and depressive-position functioning. In my view there is a clear connection between the competent functioning of an individual, essentially in a depressive position state of mind, and an institution in work group mode. Similarly, paranoid/schizoid-position functioning in the individual has much in common with a basic-assumption institutional state of functioning.

All of these concepts in turn describe and determine the behaviour of individuals in groups and institutions, and the resulting leadership "requirements", both unconscious and conscious, that then trigger the process of an individual with a suitable 'valency' to take up the leadership and to enact the requirements, both good and bad, of the position. Both Freud and Jaques have taken the story of Judith and Holophernes from the Apocrypha and shown with great clarity the interaction between unconscious group process, leadership style, and the consequences of such a dynamic.

In essence, Holophernes, the leader of the vastly superior besieging army, ran the army on the basis of charismatic leadership with the followership in a basic assumption dependency state of mind. When Judith cut his head off and displayed it to his troops it was as if they in turn had "lost their head" given their dependency state and all fled the battlefield. A sobering example of the price of charismatic leadership and dependency if ever there was one.

THE CORE PROCESS OF FOLLOWERSHIP

Followership is and must be an active participative process. It thus needs to be differentiated from a passive dependent state of mind of the individual or the group, and also from an uninvolved or denial of involvement state of mind which leads to a "boarding house" mentality in the institution. By this I mean a state of minor participation only inasmuch as it affects one's personal comfort or work, or the state of one's immediate grouping, but in which there is no responsibility acknowledged for or taken for the overall venture, nor is there any passion about it either. It is of course legitimate at times to represent the needs of and point of view of one's sectional interests; it is harder to draw the line between a sectional partisan approach and one's responsibility for the overall organisation. The risk inherent in this process is that the "overall" perspective is delegated at both a conscious and unconscious level to senior management, leadership, or the leader, resulting in a splitting process and the disowning of personal responsibility for the overall good of the organisation.

So, how does the process of active followership function? It must clearly be based on a process of consultation, participation, and involvement. It cannot, however, operate on the basis of consensus management, for that in effect means the unspoken but nevertheless true exercise of a power of veto, often on behalf of sectional interests, sometimes in the service of resistance to change. Consensus management can therefore only make decisions that are either inevitable or decisions that are of no consequence. As decision-making, in essence, is about the weighing-up of risks and the probable consequence of such decisions, it is not surprising that it is at these crucial stages that decision-making by consensus falls down and therefore, in my view, does not justify a place in the management process. On the very rare occasions where a contentious decision is reached by consensus, I believe that the outcome is often misleading because the "consensus" rapidly falls apart and then the process has to be embarked on all over again.

The creation of an arena for debate, both in the mind and in reality, is obviously a key requirement for exercising one's followership role. More contentious is how long the process of debate and consultation should take. Sometimes the latter decision is taken out of one's hands by events beyond one's control, for example deadlines might be imposed by financial factors, legislative acts, and so forth. But even then it is not unusual to have disputes about time available, with management curtailing time for discussion, presumably "fearing the worst", whilst the followership is wanting to extend the time, presumably also "fearing the worst". A senior colleague recently said to me, "You speak a lot about authority and leadership, but you seem to not believe in 'working through'"; an interesting comment reaching to the heart of the leadership/membership interface and the question of how much time is needed for *working through* and how much is too much time allocated to foot-dragging resistance. "Working through," as a basic psychoanalytic concept, implies the state of coming to terms with and accepting a psychic situation, for example the loss of a loved one. It therefore is a process of coming to terms with a loss and an acceptance that a new state of affairs prevails. A certain amount of time is required for the process—in the case of mourning traditionally about 18 months. There is of course no guarantee that after that amount of time the issue will have been worked through—it may be that the loss is as unacceptable as ever and that a near delusional state of harking back for what once was prevails.

Any change requires the giving up of something, be it a way of working or a state of self-perception, and the fact that what is being given up might have been only ambivalently valued, as it could be with an ambivalently loved or even hated person, makes no difference to the process of working through and mourning. The same of course also applies to working practices that one had mixed feelings about and that one fights to retain once they become a part of the process of managing change or part of management proposals for change. The core question is thus whether giving up something is an appropriate activity in the service of moving forward on the path of organisational change, or whether it is an inappropriate, often fashion-determined request for the giving up of something that is best retained as part of the overall institutional culture. Is resistance to giving up something then resistance to change, or is it on-task valuing of tradition? There is no easy answer to this question. No doubt, however, the question needs to be debated—robustly at times—but such a debate has to be held against the backcloth of the primary task, for it is only against that parameter that a true measure of continuity for the organisation, its members, and its products can be reached.

As a general maxim I am repeatedly struck by the fact that whereas outsiders and leaders regularly "see", speak about, and attempt to understand and to manage resistance to change, it is remarkable how the very possibility of being caught up in resistance to change is something that is never debated or acknowledged in discussions with membership. Being "too busy" to have had time for discussion or consultation, or not having had time to reach that item on the agenda is about as far a semi-acknowledgement one can get. Perhaps one should not be surprised by this; after all one would not expect a patient in psychotherapy or psychoanalysis to "own up" to being in the grip of resistance to an interpretation. And yet, viewed more positively, the very same process described above is also the process that stops omnipotent flights of fancy in leaders and helps to bring them down to the "depressive-position" reality of what can or cannot be achieved and in what timescale.

LEADERSHIP AND WORKING AT DIFFERENCES

It must follow from what has been said so far that difference of opinion, and particularly between leadership and membership, is a healthy and necessary part of institutional debate and functioning. The very same debate, however, also carries within it the seeds of destructiveness to the institution and its various component parts. How is the process of debate to be managed?

I believe that clearly bounded structures with clearly designated tasks and a spelt-out system of authority and constitution are essential for the debating process to work. The risk is that, if there are too many forums and too much reporting back, individual and structural responsibility is fudged or disowned, and decisions that have been taken, and legitimately so, are gone over again—part of a time-wasting exercise often condoned by management in the belief that "if we have everyone's fingerprints on the decision, then we'll have general acceptance". Sadly, this does not follow at all, and certainly does not make up for the foot-dragging, time-wasting, and haemorrhage of creativity.[3] The question is often raised how important structure is in the area of institutional "working through" of issues. A lack of structure makes for a process that easily succumbs to the basic assumption activity described so lucidly by Bion. The situation is even worse if it takes place in a large group (Turquet 1975). If the debate is structured and competently chaired, it has the highest chance of complying with

---

3. Inevitably there are also times when there is a stalemate in the debating process: it is logical that at times such as this expert external consultancy should be sought. Understandably consultancy has a poor record, partly because it is often called upon on a partisan basis to do management's will (as perceived). It is also not uncommon for the brief to be poorly written, or written by one side or the other, and thus partisan. A jointly written brief to investigate an issue and to report back with recommendations by a certain date seems a sensible way forward.

work-group criteria (Bion 1961). But the design of the meeting and the interconnectedness of meetings (if that is necessary given the size of the organisation) must be planned and co-ordinated, else the result is a combination of misunderstandings and a free-for-all. Bion's concept of "containment" gives a good indication of how a work-group orientated discussion of issues might be had.

On the other side there are many situations where the organisation is managed or directed with the minimum of discussion or consultation. The end result of this style is often either a "boarding house" mentality, noted previously, where there is no sense of a common venture, or alternatively a factional or tribal sort of break-up of the membership, with some being drawn into the roles of "favourites of the director" whilst others feel very undervalued.

In attempting to find structures to further dialogue and debate concerning the organisation's future, certain pitfalls appear with regularity. One is a reopening of the debate as to the primary task of the organisation. This is a tricky one, for whilst at some level going to one's sponsors, or paymasters or patients, and to say, "What do you want us to do?" is obviously part of good and realistic feedback, it is not something that can be done as an ongoing process, nor is it reasonable to expect coherent or reasonable answers in response to this question. As mentioned previously, there has to be continuity of task, as there has to be a vision and strategy for the future.

It can also be noted that in "psychologically orientated" organisations there is another classic pitfall—that of the member of staff who colludes with unconscious group processes, as well as perhaps personal psychopathology, in order to present themselves as a victim or a patient to be dealt with on a therapeutic rather than managerial basis. This individual is then offered/given a great deal of "support" while the organisational consequences are not addressed at all. The rationale for nonintervention on the organisational level often takes the form of "we all have strong and weak phases or elements in our team, and a humane organisation makes a place for all". While this is undoubtedly a valid approach that needs to be invoked from time to time, it takes no account of managers at times needing to make uncomfortable and unpopular decisions in the service of the organisational whole. The above-mentioned philosophy then comes in as a useful mechanism to let managers off the hook and help them evade and avoid painful elements of their work role, a classical example of the "rationalisation" mechanism originally described by Freud.

THE LEADER'S MEMBERSHIP AND THE MEMBERS' LEADERSHIP

In any leader/member structural dichotomy there is always the risk of split-ting and projective identification with unwanted, unacknowledged, and dis-owned aspects of the self being seen often with crystal pseudoclarity in the other, leaving one free, virtuous, misunderstood, and self-righteous. Thus there needs to be the opportunity for the leader and the leadership to engage in ordinary and everyday membership activities, and for the leader not to spend his/her time in their "private dining room" being protected by a "loyal" personal assistant. Such activities would seriously impair their capacity to encompass and conceptualise the reality of membership of the organisation. And equally members must take on leadership roles open and available to them to allow their creativity and managerial skills to develop and for them to have an opportunity to identify with the hardships and vicissitudes of management and leadership, and perhaps particularly with the loneliness inherent in the decision-making requirements of the role. Reluctance to do so would alienate them further from management and create a widening of the psychic split between management and membership.

THE VEXED QUESTION OF INSTITUTIONAL MORALE

Institutional morale and its state, whether good or bad, is a key currency in the organisational marketplace. Managers and leaders almost always proclaim it to be good, members and workers usually the opposite. It is an institutional chimera that everyone knows, but there is no objective way of describing it. There are supposedly objective ways of measuring it, but many of these, such as staff turnover, are now discredited given the tight-ness of the labour market and the difficulty of changing jobs.

From a psychoanalytical perspective, it would not be out of line with the unconscious functioning of institutions to assume that maintaining the status quo would be the equivalent of maintaining good morale, even though objectively the state could be called "living in a fool's paradise". The psychic process argument would go as follows: No change means no uncomfortable adjustments to be made, and this thus counts for the best degree of happiness/good morale available, given the circumstances. This is likely to result in an increasingly paranoid/schizoid style of institutional functioning. The uncomfortableness of change by contrast would result in poor morale, particularly as it is well-known that poor morale is a supposedly potent weapon in institutional negotiations and in slowing down or modifying change. While the above could by no means be described as a depressive position style of functioning, it is a reality that this style of functioning and thought is widespread in institutions. In my view, poor morale—or a phase of poor morale—is therefore an inevitable component of the process of change.

PERSONALITY VS. INSTITUTIONAL PROCESS

The group relations model of looking at organisational functioning would assume that any individual's behaviour in an organisational setting is likely to be acting out an unconscious group or organisational process part or role. The personal element of the transaction would be in the unconscious link between the institutional role required and the element of personal psychology/psychopathology that would make that person amenable/susceptible/vulnerable to the part required to be played—what was earlier referred to as "valency". From a management point of view the question is therefore, If you wish to influence the situation, at what point do you intervene—the individual, the organisation, or both? Again, in a classic group relations exercise such as the Leicester Conference, there is no question but that it is seen as part of the institutional process and therefore dealt with at that level.

In practice, however, it rarely turns out as simply or elegantly as outlined above—a combination of both approaches is probably necessary. A managerial intervention is required to enable mutual projections to be withdrawn and personally owned, and the individual needs to be helped to recover his/her individuality, and/or sometimes helped to leave the organisation for the process is at times too fixed to be undone, and/or helped to make a new start. Helping an individual to leave, or sometimes to be dismissed, in turn often triggers a whole institutional dynamic where former complainants refuse to stand by their complaints, and some even become supporters—a sort of in-house unconscious Stockholm Syndrome. This refers to a situation where individuals held against their own will and used as hostages in negotiations with the authorities identify with their kidnappers and act for the latter's benefit. An active institutional component of this process is often the unspoken, and at times unconscious, belief that unless this process of dealing with problems is stopped one's turn will come next. This is a dynamic of a mad spinning-out-of-control in order to stop something—an issue perhaps best caught in Schiller's *Dr. Faustus*, where the more the broomstick is chopped into pieces by the sorcerer's apprentice in order to stop the chaos, the more chaos is created and multiplied many times over. On the other hand, not dealing with the individual concerned of course also elicits a price, as the management is shown to be ineffective, weak, and unable to grasp this nettle, or any other nettle for that matter.

On a personal basis it is of course always humiliating to eat humble pie, so if one or the other dire prediction made by one or the other side does not come true, it is always tempting to claim it has come true despite evidence to the contrary, or to nudge it in the direction of coming true even though it is not innately heading that way, in order to ensure that one does not suffer a loss of face or that one has to eat humble pie. Forgiveness is not a term ever written about in management or institutional books, yet it seems to me to be a key element in helping to defuse states of begrudgingness and miffedness. Those states after all are common in family life, and it should come as no surprise that a similar dynamic can often make its appearance in institutions.

ANXIETY AS AN INSTITUTIONAL ISSUE

In his classic paper "Social Systems as Defence against Persecutory and Depressive Anxiety", Jaques (1955) built the central core of understanding the origins of anxiety in social institutions. Isabel Menzies Lyth (1960) followed this with her classic study of nurses in a teaching hospital. Between them, they covered the essential elements of work-related anxiety. Less though has been written about the anxiety of facing the "unknown known". By this I mean individuals facing tasks on or over the time/conceptual horizon; tasks that can somehow be envisaged but not emotionally grasped; tasks that can be thought about and yet not adequately conceived. It is as if what is required is beyond the make-up of the individual, and in order to master the task, psychic elements have to be summoned that cause discomfort and emotional turbulence.

The natural inclination is to reduce the "length" of one's vision—keeping one's head down as regards the future, and perhaps carrying on as if nothing had happened or was about to happen. The alternative seems to be to summon a degree of omnipotence in order to face the future, but that in turn carries with it the risks of omnipotence, being out of touch, being resistant or defended against feedback, consultation, and differing views. It does require a particular inner-world balance of omnipotence and sanity to say, "The best way to control the future is to invent it", and to get away with it without falling into a delusional state, tyranny, or depression at one's inability to control the future in line with one's wishes.

Part of the process of leading/managing all members of the institution towards the future is, of course, to have them all on board. This seemingly simple requirement is in present-day Western societies somewhat of a problem, because we espouse a multiplicity of connections as being a richness, and look down upon the, as we perceive it, blinkered allegiance of the "company man" as exemplified by, say, the Japanese. And yet it is clear that the multiplicity of roles that are part of the richness of some of our workforce also make for the existence of a multitude of escape routes for those supposedly en route to the "unknowable known" future.

CONCLUDING THOUGHTS: LEADERSHIP, FOLLOWERSHIP, AND THE CREATIVE WORKPLACE

By now it is clear that the individual member of the institution, in whatever role, has to tread a careful path between a variety of loci of power and influence, and in the process must maintain his or her individuality while yet belonging to the requisite group process. Pierre Turquet (1974), in "Leadership: The Individual and the Group", addresses this issue in its full complexity, but it is not only the members of the institution who must tread this careful pathway; it is also the leaders. However, being pre-occupied with this process can itself become a perverse form of leadership, with such pre-occupations taking up a great deal of time and thought, and providing a distraction from serious attention being paid to future directions. I (Obholzer 1995) have elsewhere addressed some of these matters of "domesticity" and how they can become a defence against creativity. By "domesticity" I mean a state of falling into concern about matters of irrelevance to one's role, often to do with organisational housekeeping tasks that have been appropriately delegated to others. It is also clear that a concentration solely on "the leadership-vision thing" can be a flight from acting upon and implementing necessary institutional change, particularly if it is painful to the manager and potentially disruptive to the institution, for example sacking a member of staff.

I would like to end by quoting Alan Kay of ATARI, one of the most innovative companies of the recent past. I quote, "The best way to control the future is to invent it". And I would then like to add—having invented it you then need to think about it along the lines outlined in this paper.

REFERENCES

Armstrong, D. (1998). *The 'Institution in the Mind': Reflections on the Relation of Psycho-Analysis to Work with Institutions*. London: Free Association Books.

Bion, W. (1961). *Experiences in Groups*. Basic Books: New York.

———— (1962). Learning from experience. *International Journal of Psychoanalysis* 43:306–310.

———— (1977). "Attacks on Linking", Second Thoughts: Selected Papers on Psychoanalysis. London: Heinemann Medical (reprinted Maresfield Reprints, London, 1984).

Carr, W. (1996). *Learning for Leadership*: The Leadership & Organization Development Journal, Vol. 17, No. 6, MCB University Press.

Freud, S. (1913). Totem and taboo. *Standard Edition* 13:1–100.

———— (1921). Group psychology and the analysis of the ego. *Standard Edition* 18: 69–134.

Jaques, E. (1955). *Social Systems As a Defence against Persecutory and Depressive Anxiety*. London: Tavistock Publications.

———— (1989). *Requisite Organization. The CEO's Guide to Creative Structure and Leadership*. Arlington, VA: Cason Hall.

Klein, M. (1959). Our adult world and its roots in infancy. In *Group Relations Reader*, 2, ed. A. D. Colman and M. H. Geller. Washington, DC: A. K. Rice Institute Series.

Menzies, I.E.P. (1960). Social systems as a defence against anxiety: an empirical study of the nursing service of a general hospital. In *The Social Engagement of Social Science*, Vol. 1: *The Socio-Psychological Perspective*, ed. E. Trist and H. Murray. London: Free Association Books, 1990.

———— (1983). Bion's contribution to thinking about groups. In *Do I Dare Disturb the Universe?*, ed. J. S. Grotstein. London: Maresfield Library.

Menzies-Lyth, I.E.P. (1990). A psychoanalytical perspective on social institutions. In *The Social Engagement of Social Science*, Vol. 1: *The Socio-Psychological Perspective*, ed. E. Trist and H. Murray. London: Free Association Books.

Miller, E. J. (1993). *From Dependency to Autonomy: Studies in Organization and Change*. London: Free Association Books.

Miller, E. J., and Rice, A. K. (1967). *Systems of Organisation*. London: Tavistock Publications.

Mosse, J. (1994). Introduction: the institutional roots of consulting to institutions. In *The Unconscious at Work: Individual and Organizational Stress in the Human Services*. London: Routledge.

Obholzer, A. (1995). Thinking Creatively in Context. Tavistock 75th anniversary paper. Unpublished.

Obholzer, A., and Roberts, V. Z. (1994). *The Unconscious at Work: Individual and Organizational Stress in the Human Services*. London: Routledge.

Rice, A. K. (1958). *Productivity and Social Organization*. New York and London: Garland Publishing, 1987.

Shapiro, E.R., and Carr, A.W. (1991). *Lost in Familiar Places: Creating New Connections between the Individual and Society*. New Haven, CT: Yale University Press.

Steiner, J. (1985). Turning a blind eye: the cover-up for Oedipus. *International Review of Psychoanalysis* 12(2):61–172.

Turquet, P. (1994). Leadership: the individual and the group. In *Group Relations Reader 2*, ed. A. D. Coleman and M. H. Geller, pp. 71–87. Washington, DC: A. K. Rice Institute, 1985.

Index

Accountability, lack of, 147, 177–178

Action, in psychoanalysis vs. role consultation, 27

Adolescent and Family Treatment and Study Center, consultation to, 109–111

Affect
creating holding environment for, 108
disconnection from, 32–33, 38–40
distribution of, 152–154

Affirmation, as benefit of groups, 103

Aggression
in large groups, 88
during organizational change, 70, 81, 138

Anger, in case study, 70–71, 74, 81–82, 86

Anxiety, 103, 117, 167
causes of, 89–90, 125–126, 169
containment of, 101–102, 105, 169

and depressive position, 138–139
and organizational change, 89–90, 133–134, 138–140, 202
in organizations, 9–10, 101–102, 135, 213
primitive, 3

Arabs, peace talks with Israel, 122–123, 128

Arendt, H., 63

Armstrong, D., 4, 9, 198

Army
authority in, 63
projections onto group, 47–48

Assessment methods, used by consultants, 9

Austen Riggs Center, case study of, 175–195

Authority, 54, 71, 163, 170
in case study, 74, 80–83
delegation of, 183–185, 188
and dependence, 46, 52–54
and dependency, 51–54
distribution of, 4, 8, 110

Printed in the United States
by Baker & Taylor Publisher Services

Printed in the United States
by Baker & Taylor Publisher Services